SEX
IN THE
BRAIN

SEX
IN THE
BRAIN

How Seizures, Strokes, Dementia, Tumors,
and Trauma Can Change Your Sex Life

AMEE BAIRD

Columbia University Press
New York

Columbia University Press
Publishers Since 1893
New York   Chichester, West Sussex
cup.columbia.edu

First published in Australia by NewSouth, an imprint of UNSW Press Ltd.
Copyright © 2020 Amee Baird
All rights reserved

A complete cataloging-in-publication record is available from the Library of Congress.
ISBN 978-0-231-19590-4 (hardback)
ISBN 978-0-231-55155-7 (e-book)
LCCN 2019034638

Design: Josephine Pajor-Markus
*Cover image: Self Reflected in Violets, 22K gilded microetching under multicolored light,
2014–2016, Greg Dunn and Brian Edwards, www.gregadunn.com*

CONTENTS

The cases discussed in this book are drawn from people who were seen by the author in her clinical practice or research, or are based on published cases in medical journals. In the case of the author's patients, their names and identifying features have been changed to protect their anonymity. Some cases are an amalgamation of published case studies, and for some of these the author has created fictional life stories to emphasise that these are not just medical subjects, but people with complex personal lives. In all cases, however, the details about neurological conditions and the associated sexual changes are entirely factual.

INTRODUCTION

There is nothing more fundamental, personal or universal to humans than sex. Without it, none of us would be here. Sexual thoughts, desires and behaviours are part of daily life for many, occurring a little or a lot, but for others they barely exist, or perhaps linger only in a whisper and a tingle of memory.

So what controls our sex drives and lives? It might feel like it's our hearts, or even our genitals, but it's our brains.

Surprisingly, there has been very little research on how our brains do this. Despite our society being saturated with sexual images, the scientific study of human sexual behaviour is very limited compared to research into other human behaviours. What we know about how our brains control our sex lives comes from extrapolating from animal research, recent brain imaging studies, and observations of how sex drive and behaviour can change when a brain is damaged due to injury or disease. It is this final area of study that is the focus of this book.

Certain parts of our brains play a crucial role in controlling our sex drives and sexual behaviour, in particular the frontal and temporal lobes and specific structures within them. If these brain regions are invaded by a tumour, or

1

deprived of oxygen during a stroke, or damaged in a motorbike crash, or disrupted by the neuropathology of dementia, your sexual behaviour could change radically: you could lose all interest in sex, you could be entirely consumed with thoughts of how to get more of it, or you could suddenly develop a new sexual interest or preference.

Sex is typically not the first thing that comes to mind when you think of life after a traumatic brain injury or stroke, or during treatment for Parkinson's disease, or in the early stages of dementia, but all these neurological conditions and many more can cause dramatic changes to our sex drives and lives. This is not a topic that is raised very often when dealing with these conditions, but it can have an exciting or devastating impact on our relationships. Doctors rarely discuss it, and patients may hesitate to bring it up when changes in the brain mean that so many other things need medical attention. But wanting too little or too much sex, or changing sexual preference, can devastate relationships and in rare cases even lead to criminal behaviour. For this reason, the sexual side effects of brain disorders need to be understood.

When I decided to commit four years to a PhD in neuropsychology, I knew I had to choose a subject that I found intriguing and would keep me eager to learn more. It had to be related to patient care and have real-life implications. I hit the jackpot when I met with my future supervisor, who ran a rehabilitation clinic to support people with epilepsy who had undergone neurosurgery. She described how some of them had reported dramatic increases in their sex drive, known as 'hypersexuality'. There were a few reported cases in the literature, but no thorough investigation of why this occurred in some patients. Was it a psychological reaction to sudden seizure freedom, or did it have something do with the part of the brain that was removed during surgery? I was hooked. I still am.

I decided to write this book when I realised it had been nearly two decades since I completed that PhD research. I had the urge to share some case histories in the hope that it would bring this topic out of the bedrooms of those who have experienced changes in their sex lives through brain injury or disease and to the attention of all of us. This book is intended for intellectually curious readers who are interested in the brain and learning how brain disorders can impact on those most personal parts of people's lives – their sex drive and behaviour. It is not an academic or scholarly work with all the answers to how your brain controls your sex life. Rather, it describes real cases of people who have experienced an alteration in their sex life due to brain injury or disease, and highlights some of the research on this topic. Some of the cases describe very rare sexual side effects of brain disorders or neurosurgeries, while others outline more typical sexual effects of neurological conditions. Fascinating, and sometimes funny or frightening, all the cases illuminate links between specific parts of the brain and sexual functions.

For those with no personal experience of a brain disorder, I hope this book will provide unique insights into the experiences of people living with brain conditions. I also hope this book provides comfort to those who have experienced sexual changes in the context of their neurological disorders, knowing that there are others who have felt the same. The patients and partners I interviewed for my research were so relieved to be able to discuss their situations, and to learn that others had experienced similar things. I can still recall a comment made by the husband of a woman who had the temporal lobe of her brain removed to relieve her of seizures. He was exhausted by her post-operative sexual requests, and incredibly – albeit jokingly – enquired whether he could have the same operation. 'Can you give me one?' he asked. 'I can't keep up!'

AS A CLINICAL NEUROPSYCHOLOGIST, I HAVE THE privilege and luxury of spending several hours with my patients. This is typically an unusual experience for most people who are used to spitting out their symptoms to a rushed GP or medical specialist in under 15 minutes. When I conduct a neuropsychological assessment, I spend up to an hour taking my patient's history – their story – with a focus on their cognitive or thinking skills, such as memory. It's a bit like being a detective of the mind. I need to figure out if there has been any change in their cognition skills and, if so, what the changes are and why they have occurred. Sometimes the 'why' is already known, such as when someone sees me after having a traumatic brain injury in a car crash. At other times, the 'why' is the crucial question. A man in his fifties is struggling to find words and losing his train of thought in conversations. His driving has become erratic and he forgets to take his medications. Is he just stressed out, or is there something more sinister, like an early-onset dementia, occurring in his brain?

If my patients come with a partner or another family member, I ask them for their views. Sometimes they are happy to talk about their loved ones in front of them; at other times they want to talk in private, sharing their worries and frustrations about this person who they feel they don't know anymore. I ask about the patient's schooling, their work and home lives, their mood and medical history. This luxury of time means that people often tell me things that they might not have shared with their other doctors.

I don't ever ask about their sex lives, but this doesn't mean they don't bring it up.

I've had a patient tell me spontaneously about how she had sex with her father as a teenager and it was 'a wonderful learning experience'; a man explain how he had learned from an alleged victim that his own father was a paedophile; and a teenager confide that while her mother thinks she is employed as a late-night cleaner, she is actually working as a prostitute in a local brothel. I've had the mother of a severely brain-injured young man explain how she comforted her son's girlfriend after he told his girlfriend he didn't like having sex with her as her 'vagina was too loose', and the carer of an elderly man tell me that since his brain injury he had begun discussing his previous sex life, particularly his love of 'putting fingers up bums'. Usually this information is a side story to what is most relevant to my cognitive detective work, but sometimes it is not. Sometimes it is the most prominent change in the context of a brain injury or disease.

Each chapter of this book includes personal accounts from individuals who I have met in my clinical practice or research, or who have been described in the scientific literature. Through these case studies, I will describe what parts of the brain are implicated in sexual behaviours, what sex can actually do to our brains, and how love can alter and be altered by our brains. You will hear about the sexual side effects of neurosurgery, and rare cases of paraphilias – sexual arousal or interest in atypical objects or activities – that can manifest in people with brain injuries or disease, and how such cases raise complex questions for the criminal justice system.

When I decided to become a clinical neuropsychologist, I certainly didn't expect that I'd end up writing about sex, but I continue to be completely intrigued and intoxicated by this topic. I hope you are too.

1

YOUR SEXIEST
BRAIN BITS

Like the Bloodhound Gang song 'The Bad Touch' states, we are indeed mammals.

In his book *Why is Sex Fun?* Jared Diamond calls his opening chapter 'The animal with the weirdest sex life'. That animal is the human. We *are* mammals, and a lot of what we have learned about how our brain controls our sex lives comes from research done on animals, typically rodents and small mammals. But there are significant differences between animal and human sexuality, so extrapolating from the sex life of animals and applying the findings to humans is problematic.

In regard to our sex lives, says Diamond, 'we are the ones who are bizarre' when compared to the thousands of other mammalian species on the planet. He provides many examples of our peculiarity. Humans, for example, do not just have sex to reproduce but have sex for fun, at any time, typically in private and often with the same partner over a long period of time. In contrast, other mammals have sex for the sole purpose of reproduction, at specific times only, typically in public and with many partners. Animal sexuality is largely dependent on hormonal mechanisms, while human sexuality involves a myriad factors in addition to hormones, including mood and personality.

Yet even though the use of animal models is obviously limited, there is a relationship between particular brain regions implicated in animal research and observations in humans. One brain region that will feature throughout this book is the temporal lobe; you have two of them, one on each side or hemisphere of your brain, behind your ears. Within the temporal lobe is a structure called the amygdala (pronounced *ahMIGdala*; plural *amygdalae*), which could arguably be referred to as your sexiest brain bit. By 'sexiest', I mean the part of the brain that has most commonly been associated with changes in sex drive and behaviour.

Over 130 years ago, in 1888, scientists Sanger Brown and Edward Schäfer were conducting experiments on rhesus monkeys. They removed both temporal lobes of these monkeys in an operation called a 'bilateral temporal lobectomy'. In their reports about how the monkeys behaved after surgery, they made a brief reference to one female monkey who showed 'uncontrollable passion on the approach of other monkeys, so that it is now necessary to shut her up in a cage by herself'. Fifty years later, German psychologist Heinrich Klüver was performing behavioural experiments on monkeys, and noted that when they were given the psychoactive compound mescal they made chewing and licking movements and had convulsions; similar symptoms had been described in humans who had epilepsy, specifically temporal lobe seizures – or seizures that arise from atypical activity of the nerve cells in the temporal lobe. Klüver then teamed up with American neurosurgeon Paul Bucy to perform temporal lobe resections (that is, removal of parts of the temporal lobe) of rhesus monkeys, and they unexpectedly discovered an intriguing syndrome that now bears their names: Klüver-Bucy syndrome. The symptoms of the syndrome are visual agnosia (an inability to recognise objects by sight), hyperorality (a tendency to examine all objects by mouth), hypermetamorphosis (an irresistible impulse to react and attend to visual stimuli), emotional changes (an absence of fear and anger, often referred to as the 'taming effect'), changes in dietary habits and – you guessed it – hypersexuality (a dramatic increase in sex drive).

This research was the first to demonstrate that the temporal lobes are involved in controlling sexual behaviour. The monkeys manifested hypersexuality about three to six weeks after they had undergone bilateral temporal lobectomy, and it involved both a quantitative increase and qualitative change in sexual behaviour. They engaged in indiscriminate mounting of animals of the same and the opposite sex, and of inanimate objects. They had sexual intercourse for up to half an hour, only to mount again immediately after dismounting, as though they had overdosed on Viagra. Males lifted up other males by their erect penises. They orally and manually explored their genitals while in all kinds of positions, for example biting their own penises while suspending themselves upside down from the top of their cage and swinging back and forth. In other words, they wanted sex all the time, and in any way they could get it. These behaviours were not seen in monkeys that had undergone unilateral (one-sided) temporal lobectomy, or in monkeys that had not been operated on at all.

Subsequent reports of partial or complete Klüver-Bucy syndrome in humans then confirmed the temporal lobe's role as one of our brain's most important regions for sexual behaviour. The first published case of the syndrome in humans appeared in 1955. A 19-year-old man had undergone a bilateral temporal lobectomy to treat his temporal lobe epilepsy. After the surgery he showed all the symptoms of Klüver-Bucy syndrome except hyperorality. His hypersexuality was expressed through sexual disinhibition, increased masturbation and out-of-character homosexual tendencies. He also had a dense amnesia (memory loss). It's surprising that he underwent this surgery in the first place; perhaps his surgeon had not yet heard of the devastating case of Henry Molaison, better known as 'HM'. HM had undergone the same operation two years earlier to treat his severe epilepsy and had been left with a global amnesia. He could not learn new things and also could not recall the 11 years before his surgery; if you talked to him, he would forget you and the conversation as soon as you left the room. HM is undoubtedly the most famous neuropsychological case study of all time, and he taught us a lot about memory, in particular how the hippocampi (singular: hippocampus) – two seahorse-shaped structures, one within each temporal lobe – are crucial for making memories. There are no reports that HM had Klüver-Bucy syndrome, despite his surgery including resection of his amygdalae.

Twenty years later, the first report of full-blown human Klüver-Bucy syndrome was published. A 20-year-old man who had damage to both his temporal lobes due to herpes encephalitis, a rare type of brain inflammation, showed visual agnosia and prosopagnosia (inability to recognise faces) instead of the 'psychic blindness' observed in Klüver's monkeys, and placidity and flattened affect – that is, a general lack of emotional expression – as opposed to the monkeys' loss of fear and anger. In this case, the subject's sexual orientation changed from a heterosexual to a homosexual preference. This is one common feature of the hypersexuality described in human Klüver-Bucy syndrome; the other typical feature is sexual disinhibition. For example, in addition to all the other symptoms of the syndrome, a 32-year-old woman with herpes encephalitis had no interest in having sex with her husband but made sexual advances, both 'manually and orally', towards female hospital staff. A 57-year-old man who had a severe traumatic brain injury in a car accident made indiscriminate sexual advances towards male and female staff in hospital, and a 52-year-old lawyer who developed seizures showed no overt hypersexuality but rubbed his genitals so frequently that he developed an abrasion on the shaft of his penis.

Human Klüver-Bucy syndrome has been reported in people with a range of neurological conditions, including herpes encephalitis, temporal lobe epilepsy, stroke affecting bilateral temporal lobes, and Alzheimer's dementia. What all of these conditions have in common is that they involve the temporal lobes. Every human diagnosed with the syndrome has bilateral temporal damage, and in most cases the damage includes the amygdala. This is certainly the case if hypersexuality is a feature, and echoes what has been found in animal research. Following on from Klüver and Bucy's research, further studies of primates, cats and rodents focused on the amygdala; destruction of both amygdalae in these animals revealed the same sexual changes that had been observed in rhesus monkeys who had been given bilateral temporal lobectomies. Although the temporal lobe is widely considered to be a critical part of the brain in controlling our sex drives and behaviours, it is the amygdala that takes the award for the structure within the temporal lobe that is most commonly implicated in cases of altered sexual behaviour via brain injury or disease.

In neuroanatomy textbooks, the amygdala is always referred to as an 'almond-shaped' structure. This analogy reminds me of my neuroanatomy class which was scheduled just before lunch. I always felt hungry while dissecting my assigned human brain, and although I was in awe of this firm jelly thing that was responsible for everything the person had ever felt, remembered or dreamed of, I couldn't shake the thought that it looked like something you could throw into a stir-fry. The amygdala is found in the middle of each temporal lobe, so we have two amygdalae – one in our right temporal lobe and one in our left. It's about the size of a five-cent piece. It plays a crucial role in processing our emotions, and abuts the hippocampus, which is critical for memory. We know that memories and emotions are closely intertwined, so it makes sense that these two structures are neighbours.

Thirteen different nuclei make up the amygdala, and each nucleus interconnects with different brain regions, including the olfactory bulbs (involved in processing smells) and the hypothalamus. The hypothalamus, in turn, regulates our endocrine system (hormones) and our autonomic nervous system (bodily functions such as heart rate and breathing). Each amygdala is also connected to other brain regions, such as the frontal lobes, including the medial, orbital and prefrontal regions. These rich interconnections are important to bear in mind when we consider the amygdala's role in sexual behaviour. It does not act alone; rather, it is part of a wider 'sexual neural network'. Other brain structures and regions that are part of this network will feature throughout this book.

FOR MY PHD RESEARCH, I SPENT FOUR YEARS INVESTIgating changes in sex drive and behaviour after epilepsy surgery. My supervisor was in charge of a service that offered pre- and post-operative counselling to patients who were undergoing neurosurgery for epilepsy, and she had found that some had reported hypersexuality after their operations.

These patients told her their sex drives had increased dramatically, and some had even confided that they had experienced a change in sexual orientation. In some cases, the partners of patients had confirmed this surprising surgical outcome (see Chapter 3). The purpose of my PhD was to explore why this was happening. Was it because they were having fewer seizures, so they were more confident, more 'in the mood' for sex? Or was it to do with the part of the brain that had been removed? Was it perhaps a combination of both these reasons?

We already knew about cases of hypersexuality as part of human Klüver-Bucy syndrome, so there was no doubt about the potential role of the temporal lobe. I interviewed over 70 people who had undergone neurosurgery for epilepsy to find out about their sex lives. I also looked at their brain scans before they had surgery. I was particularly focused on their amygdalae. I wanted to know if the size of their amygdalae had any impact on what happened to their sex lives after surgery. I spent months of my life staring at the brain scans of these patients, locating their amygdalae and literally colouring them in (with numerous clicks of the computer mouse) to then determine their size. This can now be done in a more automatic way, but back in the early 2000s when I was doing this work, it was laborious. There were no automatic computer algorithms for determining the volume of various brain structures. I had to carefully identify the structure on each image on which it appeared, then painstakingly colour it in until there was no sign of it. Then I could tap in the code to add up those coloured pixels and the computer would spit out a number that represented the calculated volume.

My supervisor for the neuroimaging part of my thesis had a team of 'neuro nerds' (neurologists, neuropsychologists, neuroscientists) in his lab, all doing different research projects for various degrees. The lab was in the basement of the hospital; my supervisor jokingly referred to it as the *laboratoire*, but it was far from the luxurious abode that this pronunciation suggests. I referred to it as 'the dungeon', which I think was a more accurate description. It was windowless, cold and bare – just desks and computers, and people staring at screens covered in images of brains. My supervisor had a favourite saying – 'Life is short' – which only exacerbated my desperation to escape the dungeon and get on with what I felt was 'real' work: seeing patients. Spending hours staring at brain scans in the dungeon turned me off ever wanting to do neuroimaging research again.

Nevertheless, I found the idea that there could be a link between a patient's amygdala size and sexual outcome after surgery fascinating, so I persisted with my amygdala painting for many months. In addition to all the patients' brain scans, I also had to measure the amygdalae of healthy people as a comparison 'control' group. I had 45 patients and 46 healthy people, so 91 brain scans to review, with two amygdalae to measure on each, making a grand total of 182 amygdalae. Initially, each one took hours, but I got quicker at it over time. In the end, all that amygdalae painting paid off. When I did my statistical analyses, I was surprised to find that I had actually found something.

There was a significant relationship between the size of the amygdala in the healthy temporal lobe of patients – that is, the opposite or 'contralateral' lobe to the one where the seizures were coming from – and the person's sex drive after they had a unilateral temporal lobectomy. The bigger the amygdala in the healthy temporal lobe, the better the person's sexual outcome. There was also a significant positive relationship between the volume of the contralateral amygdala and the maximum degree of sexual change. Patients who reported a post-operative increase in sex drive had a significantly larger amygdala in their healthy temporal lobe when compared to patients who reported a sexual decrease or no change, and when compared with the control group of healthy people. These findings provided the first evidence ever recorded for an association between amygdala size and human sexual behaviour. When it comes to amygdala size and sexual outcome after temporal lobectomy, bigger *is* better.

So why would a bigger amygdala be better? One possibility is that the amygdala, through its extensive connections with the hypothalamus, plays an inhibitory role – in other words, it puts the brakes on – in the regulation of the endocrine or hormonal mechanisms of sexual behaviour. Removing the temporal lobe, including the amygdala, on the side where the seizures come from gets rid of this inhibition (brakes off) on the hypothalamus, which leads to a post-operative increase in sexual behaviour. The connections between the remaining amygdala and hypothalamus could also account for variations in sexual outcome.

An alternative explanation is based on the amygdala's role in emotional processing. It is thought the amygdala regulates the attachment of emotional significance to things and applies reinforcing or discriminative properties to sensory stimuli. In other words, it helps us decide what or who we should or shouldn't approach, feel attracted to or run away from, love or hate. For example, if you were walking home in the dark and were approached by a tall, aggressive-looking person asking for money, your amygdala would be responsible for alerting you to the potential dangers of this situation, and kickstart your emotional (fear) and behavioural (fight-or-flight) responses. So, according to this proposal, amygdala damage does not disrupt a specific sexual/endocrinological mechanism. Rather, it disturbs the emotional processing of stimuli, which then causes inappropriate and indiscriminate responses. This theory suggests that rather than playing an inhibitory role, the amygdala has a positive effect on sexual behaviour by allowing the appropriate attachment of emotional significance to sexual cues. A larger amygdala may function better in its role in processing emotional and specifically sexual information and the attachment of significance to it, which would increase the likelihood of a sexual response, resulting in an increased sex drive. A bigger amygdala would help you to notice and feel aroused by a sexual cue and make you more likely to accept that cue and go for it.

What about the cases of hypersexuality after temporal lobe surgery, when some patients' partners complained that they couldn't keep up with their loved one's newfound post-surgical sexual desires? Could that be human Klüver-Bucy syndrome? I don't think so. The sexual increase described by the patients I interviewed was qualitatively different from the reports of hypersexuality in Klüver-Bucy syndrome in both animals and humans. Hypersexual animals who have bilateral amygdala damage show indiscriminate sexual behaviour towards inappropriate objects; for example, they will attempt to mount inanimate objects. Similarly, hypersexual changes in human Klüver-Bucy syndrome involve indiscriminate sexual behaviours, ranging from sexual advances (verbal or physical) towards strangers to homosexual advances that were previously uncharacteristic in the patient. Sexual movements such as hip thrusting, and exhibitionistic behaviours such as disrobing and public masturbation have also been described. These behaviours represent a decrease in selectivity of the target of sexual advances, and of the time and place of sexual expression.

In contrast, the post-operative sexual increase described by the patients who had undergone temporal lobectomy mostly involved an increase in sex drive *without* indiscriminate sexual behaviour. Using the term 'hypersexuality' in relation to Klüver-Bucy syndrome confuses these two kinds of sexual behaviour – *increased* sex drive and *indiscriminate* sexual behaviour – and my research suggests that these are two distinct outcomes that may be caused by distinct neurological mechanisms. The fact that I found a positive relationship between amygdala size and sexual outcome also supports the notion that the patients I studied did not have Klüver-Bucy syndrome; if they did, I would have found a *smaller* dysfunctional amygdala associated with increased sexual behaviour.

'Bigger is better' was the title of one of the talks I gave about my PhD research. It was fun thinking of titles. It was around the time that *Sex and the City* was a huge hit, so 'Sex and the single amygdala' was another one I used. 'The amygdala and sexual outcome after epilepsy surgery: Does size matter?' was the title of the research study I presented at my first overseas conference in the United States. It certainly caught people's attention (see Chapter 3). But after four years of thinking and writing about a single topic, and many months in 'the dungeon', as soon as the PhD was done I yearned to run away to a new life – and that's exactly what I did. I wanted to work as a clinician and explore the world, so I moved to London. I was lucky to get a job at the National Hospital for Neurology and Neurosurgery, a world-renowned specialist hospital; I was thrilled to find out that many famous neurologists had worked there. My 'sexy' PhD faded into the background as I was trained in the busy life of a clinical neuropsychologist. Until I met a patient who brought it all back …

2

'GIVE IT TO ME BABY' OR 'NOT TONIGHT, DARLING'

Jack was my final patient for the day. I groaned when I saw that he was an inpatient. That meant lugging all my testing equipment up four flights of stairs, then competing with the cacophony of squeaking meal trolleys, patients crying out in pain and beeping medical equipment as I did my assessment. The smell of the hospital wards, with their toxic mix of cleaning liquid, bodily fluids and hospital food, always made me feel queasy.

I checked the referral note again. This handwritten scrawl was more detailed than usual. The medical doctors usually only wrote the name of the patient, their neurological condition and 'neuropsychology assessment' – or 'psychometry', a term that made me cringe, as it made me feel like a testing machine. This referral had a 'please', which was rare. It was also full of acronyms: 'TBI due to fall from scaffolding. GCS at scene 6. MRI brain R > L frontal damage. Disinhibited. Please assess.'

To explain:

TBI = Traumatic brain injury, or a brain injury caused by a trauma, in this case a fall.

GCS = Glasgow Coma Scale score. This is a measure of the severity of the brain injury, a score out of 15, with lower scores indicating a more severe injury.

MRI = Magnetic resonance imaging. This is a type of brain scan that gives a detailed image of brain structure.

R > L = Right greater than left.

I found the head nurse on the ward and asked where I could find this 'disinhibited' patient. She gestured down the corridor and chuckled. 'You'll have fun with him,' she said in a thick cockney accent as she turned away.

He was lying down staring at the ceiling when I arrived at his bedside. I introduced myself and he tilted his head slightly towards me.

'Oh, thanks for coming to see me,' he began. 'Can you just give me a quick blow job? Just a real quick one. I promise I haven't been with any other women. I smell fine. Just make it real quick – no one will notice. Go on, I really want you to suck it, real quick. I won't take long.'

I laughed, blushed and tried to distract him with my tasks, but he was unstoppable. I was curious to see how long he would persist with his sexual requests. I usually spent at least a couple of hours with my patients, checking all their thinking skills including memory, IQ and attention, but his verbal advances were relentless. He had no interest in my cognitive tasks. His mind was fixated on one thing only: sex. I gave up after half an hour. I wondered how his wife was handling this unexpected outcome of his fall.

A decade later, back in my home country on the other side of the world, in the comfort of my private rooms I saw Tessa, an 18-year-old woman who had suffered a traumatic brain injury in a road accident. She had been in the front passenger seat of a car driven by her boyfriend; he had been drunk and speeding after a night out celebrating the end of high school. Tessa had suffered only minor physical injuries, but her mother explained that this was both a blessing and a curse: everyone thought that since Tessa looked the same, she *was* the same. The invisibility of her brain injury made it difficult for her friends to understand the changes in her behaviour.

Tessa draped herself across the armchair in my room, her model-length legs slung over one arm of the chair, her head resting on her mother's shoulder. She twisted the ends of her long hair around her fingertips for the entire interview, her other fingers alternating between her lap and her mother's hand. She let her mother do most of the talking, occasionally contributing in a soft and childish voice. She seemed like a shy eight-year-old girl in a woman's body.

Her mother gave me her daughter's history, then hugged her daughter goodbye as she stepped out of the room. Then I asked Tessa to move closer to the desk so we could start the assessment. She swung her legs back over to the front of the chair in a dramatic sweep, flipped her hair back over her shoulder and pulled the chair up to the desk. I asked if there was anything she wanted to discuss now that her mother had gone. She leaned forward and smiled.

'Actually, there is *one* thing,' she said. Her eyes widened and she suddenly appeared more mature and assertive. 'Since my accident, I feel like I want to fuck every guy I meet.'

I HAVE WORKED AS A CLINICAL NEUROPSYCHOLOGIST for nearly two decades and met thousands of patients with all sorts of brain injuries and diseases. I typically spend a few hours with each of them. The first part is about an hour of getting to know their story – who they were before their brain changed, and how they feel their neurological condition has affected them. This is my favourite part of the assessment process. I have the luxury of time, and the experience for the patient is different from what they go through in other medical appointments where they have a few brief minutes, rather than hours, to share their story. I think this is why I hear things from these patients that they may have never discussed with their other doctors: their secrets, their wishes, their regrets. I then do a couple of hours of cognitive or thinking tasks, to check on their memory and other cognitive functions.

It's impossible to remember every patient, but some are hard to forget. It's not part of my routine practice to ask about people's sex lives, but sometimes patients or their partners will spontaneously raise the subject. And sometimes, as in the case of the two patients I just described, it is the most memorable feature of their presentation.

Jack and Tessa both had traumatic brain injuries and experienced hypersexuality. Why? In both cases it was a result of *where* their injuries had damaged their brains. Traumatic brain injuries typically occur in motor vehicle accidents. People hit their heads on the windscreen or, even worse, are flung through it headfirst. The frontal and temporal lobes of their brains are the most vulnerable regions, given their positions in the brain, and these two regions are critical parts of the sexual neural network (see Chapter 1). Any type of brain injury or disease that affects these brain regions or specific structures within them, such as the amygdala, can alter the functioning of the sexual neural network and cause changes in sex drive and behaviour. This is typically an unexpected surprise, especially for the person's partner.

Traumatic brain injuries are not the only type of neurological condition that can cause hypersexuality. Certain types of dementia can also give rise to this sexual change, as in the case of Terence. He was a highly successful lawyer, a partner in a big law firm. Polite and introverted, he loved his job and planned to work until he was too frail to walk up the two flights of stairs to his office. But his retirement came much earlier than expected. In his mid-fifties, over the course of a year, he transformed into someone who his colleagues, clients and family did not recognise. He put on ten kilograms after suddenly developing a sweet tooth; the cleaning staff at his office would find lolly wrappers scattered under his desk and bags of sweets protruding from his filing cabinet drawers. He became impulsive and careless in his work, and rude to clients. He made inappropriate jokes to his colleagues and showed no insight into how his behaviour was affecting others.

After numerous complaints from his secretary, and clients requesting to change lawyers, his old high school friend and partner in the firm ended up calling Terence's wife. She burst into tears and said that he was 'no longer the man I married'. She confided that he was demanding sex multiple times a day, and if she didn't comply he became verbally aggressive or stormed off to his study to watch porn for hours. After that phone call, Terence's wife insisted that Terence visit his GP with her, and it didn't take long for him to be referred to a neurologist. He was eventually diagnosed with behavioural-variant frontotemporal dementia.

Frontotemporal dementia is a rarer form of dementia than Alzheimer's dementia, and has a younger age of onset. There are different subtypes; some mainly affect language functions, resulting in difficulties with verbal expression, while the behavioural variant primarily affects a person's personality and behaviour. This is unlike Alzheimer's dementia, in which the main symptom is memory difficulties. The behavioural changes are often first noticed in social situations when a person shows a lack of empathy, or may become disinhibited and inappropriate. As the name suggests, this type of dementia affects the frontal and temporal lobes of the brain. Your left and right frontal lobes sit behind your forehead. They are the last parts of the brain to mature, and are the largest of all four lobes of the brain. They are extensively connected with other cortical (top surface) and subcortical (inner, deeper) parts of the brain. The frontal lobes comprise three major regions: (1) the primary motor cortex; (2) the premotor and supplementary motor cortices – both these regions are involved in controlling voluntary movement and integrating sensory and motor information to perform actions – and (3) the prefrontal cortex. This prefrontal area can be further subdivided into numerous regions, including the superior medial cortex, controlling 'energisation' or the initiation and sustaining of behaviours; the orbitofrontal cortex, mediating social and emotional behaviours; and the dorsolateral prefrontal cortex, involved in task setting and monitoring, among other functions.

The frontal lobes are often referred to as the 'executive' or 'conductor' of the brain, as they control many social, behavioural, emotional and cognitive functions that are considered to be the very 'essence' of what makes us human. To name a few, they enable us to plan, organise, make decisions and act on things. They help us regulate and control our emotions, are involved in motivation and reward processes, and control an array of cognitive functions including attention, memory, problem solving and language. They make us who we are in that they play a role in personality and complex social and emotional behaviours, such as empathy, humour, deception and creativity. Entire books and PhD theses have been written on frontal lobe function, so it would be impossible to cover it all here. For our purposes, the frontal lobe's critical role in human sexual behaviours is most relevant. Everything sexual that humans do involves a complex interplay of motor, social, cognitive and emotional functions, so it makes sense that the frontal lobes play a critical role in the sexual neural network. For example, in the broadest sense, our frontal lobes (specifically the regions that mediate motor function) control the movements we need to make to have sex, inhibit us from having or seeking sexual responses at socially inappropriate times, and are involved in the myriad social and emotional functions that accompany sex, such as the ability to empathise with our partner.

But *how* do we know this? The development of brain imaging techniques has enabled major advances in the scientific study of the neurological control of human sexual arousal. As well as structural brain imaging methods that reveal the structure of the brain, there are now 'functional' methodologies too. These measure the function of the brain through blood flow (such as fMRI – functional magnetic resonance imaging) and glucose metabolism (such as PET – positron emission tomography). What we do know so far about what happens in healthy brains during sexual arousal has primarily come from studies in which people look at both sexual images and non-sexual images while their brains are scanned. A review and meta-analysis of these studies found that various frontal regions (premotor cortex, orbitofrontal, medial prefrontal and anterior cingulate regions) are consistently activated, along with other brain regions such as the famous amygdala, the hypothalamus, the thalamus and the substantia nigra – all of which will appear later in this book. Heterosexual and gay men showed similar patterns of brain activation, as did women and men. The topic of how the healthy brain controls sexual arousal and behaviour warrants its own book, but for our purposes the message is clear: the research confirms that the sexual neural network consists of many different regions spread throughout the brain. Disruption of any of them in brain injury or disease can result in an altered sex life.

So: back to the frontal lobe and its role in the sexual neural network. In rare cases of behavioural-variant frontotemporal dementia – like Terence's – hypersexuality can be the first or 'presenting' symptom. However, the typical sexual change in this type of dementia is not hypersexuality but *hypo*sexuality, or a lack of interest in sex. People with this condition will fail to initiate any form of sexual behaviour, even simple affectionate gestures such as holding hands or offering a hug. If their partners try to initiate these gestures, they are likely to hear, 'Not tonight, darling'. A person with frontotemporal dementia will typically be oblivious to the impact of their hyposexuality on their relationship.

This obliviousness to a sexual change has also been observed in people with acquired brain injury. Sally had been complaining of blurry vision, but she and her husband just thought she needed glasses and joked that she was 'past her prime' now that she had hit her mid-forties. Her GP sent her for a brain scan, just to exclude anything 'sinister'. A mandarin-sized tumour was discovered deep within her brain. It had to come out, and she was immediately admitted to a neurosurgical ward. During her surgery, Sally had a stroke that affected a crucial brain structure called the thalamus, which has rich interconnections with all the brain regions – in particular the frontal lobes.

When she emerged from the intense fatigue that engulfed her in the first six months after the stroke, only her closest friends and immediate family realised that she was different. Everyone else commented how lucky Sally's husband was to have Sally back – alive, walking, talking. She looked normal, but they had no idea that she was not the wife he knew. 'My wife went into surgery and never came back,' Sally's husband said. 'It is like having another child.' Sally had always been witty, but now her jokes were childish. She kept referring to farts and poo. She made loud comments about 'fat' and 'ugly' people while out in public. She threw food around the dinner table while their three children watched, wide-eyed, unsure whether to laugh and join in or tell her to stop. When she caught a glimpse of her husband in the shower one morning, Sally pointed and started giggling: 'I saw your doodle!' They were never intimate again. Sally had no interest in sex, and neither did her husband now that he had suddenly become a carer rather than a partner.

It was a similar story for Rita, whose husband of more than 30 years had enjoyed a few too many beers one evening and tumbled headfirst down the stairs at the pub. When Rita arrived at the hospital she was expecting to see a broken leg, and was shocked to find her husband unconscious. The doctors asked her to sit down and looked serious as they pointed to dark patches on his brain scan that glowed on the computer screen. The only words she heard and remembered were 'frontal lobes' and 'severe'. She couldn't comprehend how a fall down some stairs could be so dramatic and cause so much damage. When I met them several years later, Rita had gotten used to her new husband, who exploded over minor irritations, repeated himself in conversation and couldn't stand her talking to him if he was doing something else. She spontaneously referred to their sex life and said, 'I've tried and there's nothing. He's not interested since his fall. That's OK. It doesn't bother me. At least he's not aggressive.' Then she shrugged and changed the subject.

HYPOSEXUALITY CAN ALSO OCCUR IN PEOPLE WITH temporal lobe epilepsy. Their lack of interest in sex develops gradually over time as frequent seizures disrupt the temporal lobe and sexual neural network. As I had learned in my research, neurosurgery that removed the temporal lobe causing the seizures could result in a dramatic increase in sex drive, and in some people this had occurred after a decades-long absence of libido. Hypersexuality in the context of temporal lobe epilepsy can be associated with other striking changes in sexual behaviour, as noted earlier – such as a change in sexual preference, or even the development of a 'paraphilia' or sexual disorder, such as paedophilia (see Chapter 7). In some cases, this has led to criminal behaviour and legal dilemmas (see Chapter 10).

Temporal lobe seizures can be triggered by an orgasm, or even cause orgasms. Seizures that are induced by orgasm are rare but they can be frightening for partners and have a significant impact on a person's sex life. They can lead to a life spent avoiding sex and fear of orgasm, which can have a devastating effect on relationships. In one case, the husband of a woman who experienced orgasm-induced seizures was so frustrated by their sex life that he threatened divorce if surgery to cure her seizures was not successful. (See Chapter 3 for a related discussion on the sexual side effects of seizure surgery.) Orgasm-induced seizures occur much more commonly in women than men and are usually associated with a right temporal lobe seizure focus. Brain imaging studies of healthy men and women have found that orgasm, and its lead-up, is predominantly associated with activation (and, in some earlier studies, deactivation) in the temporal and frontal brain regions, including the amygdala and orbitofrontal cortex; other regions involved in sensory, motor and reward processes are also implicated. It appears that if the neurons (the nerve cells) in those very brain regions are highly sensitive, perhaps due to scar tissue or other causes of seizures, such as hippocampal sclerosis (discussed in the next chapter), then a seizure can be triggered by the activation or stimulation of those exact regions that occurs during orgasm. One brain imaging study of women suggests that just imagining or thinking about genital stimulation can activate the same sensory and reward brain regions that actual genital stimulation and orgasms do.

In contrast to orgasm-induced seizures, seizures that *result* in orgasms may be savoured by those who experience them. Orgasmic 'auras' (a feeling or 'warning sign' that a seizure is about to happen) linked to seizures are also more common in women and typically arise from the right temporal lobe. Spontaneous orgasms might sound like fun, but these sexual seizures can occur suddenly and in unexpected situations. Imagine travelling on a bus during peak hour on your way to work, standing in the aisle jammed in between other passengers, and suddenly feeling a wave of tingling. You know what is coming, and you know that you are about to experience it in front of an audience. Case studies of women who experience these pleasurable seizures have found that they often keep them a secret from their doctors – for decades in some cases – even when they are undergoing investigations for epilepsy and know that orgasmic auras are part of their seizures. Despite the fact that the pleasure progresses into a seizure, some people have refused to have surgery to cure their seizures out of fear of losing these unexpected orgasms.

Apart from orgasm, there are other sexual behaviours that can occur during a seizure. Sexual 'automatisms' (automatic behaviours that the person later has no memory of) include writhing, thrusting, rhythmic movement of the pelvis and legs, and rhythmic handling of genitals or masturbation. These are rare and occur relatively equally in men and women who experience frontal lobe seizures. Sexual 'ictal' manifestations (that is, those that occur during a seizure) have also been reported, such as erotic feelings, genital sensations and sexual desire; these have been found to occur most commonly in women with right temporal lobe seizures. So there are many different types of sexual behaviours that can be associated with seizures, and these occur when the seizure focus is in the temporal (typically right-sided) or the frontal lobes.

Epilepsy, however, is not the only neurological condition that can be associated with sexual changes. Parkinson's disease is another that will come up repeatedly throughout this book. Parkinson's is caused by the death of neurons that produce a neurotransmitter – a chemical that communicates information between neurons – called dopamine. These neurons are found deep in the brain, in a region called the substantia nigra, which is part of the basal ganglia. This region has extensive connections to the frontal brain regions, which as we know are integral parts of the sexual neural network. Dopamine is crucial in the regulation of movement, but is also involved in many other functions including emotion and reward processing. (The 'reward system' is a group of brain structures and neural pathways that are activated by rewarding or reinforcing things, such as addictive drugs and sex.) Therefore a lack of dopamine can cause problems in these areas, such as depression, addiction and, in the case of Parkinson's disease, movement difficulties. The treatments for Parkinson's disease are drugs that increase dopamine production.

The first thing Steven noticed years before he developed a trembling hand and shuffling feet was that his voice became husky and soft. He struggled to sing in his church choir and couldn't talk as loudly as he used to. At his daughter's wedding the microphone cut out during his father-of-the-bride speech, and he could not project his voice beyond the front row of tables. A month later his legs started to feel heavy, and they wouldn't move as quickly as he wanted them to. It was when his right hand started trembling that he decided it was time to talk to his doctor. He didn't want to worry his wife or daughters, and reassured himself that since none of them had made any comments they probably hadn't noticed anything. But when his GP mentioned the words 'Parkinson's disease', he suddenly wished someone was there with him.

Reduced levels of dopamine in the brain cause many symptoms. The classic symptoms of Parkinson's are the motor ones, including limb tremors, a shuffling gait and slow movements. There are also many 'non-motor' symptoms that can occur, including depression and anxiety, and the striking 'impulse control disorders'. These include compulsive gambling, shopping and eating, and hypersexuality; these can manifest as part of the disease's progression or in response to treatments. Drugs that boost dopamine levels in the brain, and the neurosurgical treatment of deep brain stimulation that is used in more severe cases, are highly effective at easing the motor symptoms but can trigger compulsive behaviours that cause financial and personal havoc for patients and their families (see Chapter 4). Dopamine drug therapy is the first port of call to treat the motor symptoms of Parkinson's disease, but it requires a careful juggling of timing and dosage that can take months to sort out.

Steven started dopamine treatment and was surprised at how quickly his legs and hand responded; the shaking and shuffling subsided and he regained his smooth stride. He was relieved that his body was working in a predictable way again, but there was a catch. An unusual side effect had appeared in the first week he had started the medication: he craved sex. Previously satisfied with the once-a-week routine that he and his wife had fallen into after decades of marriage, Steven quickly found that this didn't cut it anymore. He needed sex – absolutely *had* to have it – multiple times a day. If his wife didn't oblige he would disappear into his study and masturbate. If that didn't ease the craving, he'd google porn websites. Even more surprising to him was that he was now getting aroused by men, a novel experience for him. He started frequenting gay beats. When his wife discovered a flyer for a gay bar in one of his trouser pockets, she started googling two things: the link between Parkinson's and sex, and how to get a divorce. She was shocked to find similar stories online about the sexual side effects of the drug Steven was taking. She raced upstairs to find his medication and searched the fine print for warnings, but there was nothing. She followed up her googling with two phone calls: one to their doctor, and one to a lawyer.

In December 2014, the pharmaceutical company Pfizer agreed to settle a class action brought by 160 Australian patients who had suddenly developed hypersexuality or pathological gambling after taking the drug Cabaser, a dopamine treatment for Parkinson's disease. Steven's compulsive sexual behaviour was devastating for his wife, but when she heard that a man who developed similar side effects had ended up on the sex offenders register and had to ask permission to visit his grandchildren, she was grateful that his desires had remained with adults. This is not always the case. Disruptions to the sexual neural network can result in a change to sexual preferences, not just from heterosexual to homosexual desire or vice versa, but from adult to child – with devastating consequences for everyone (see Chapter 10).

The people described here manifested extreme changes in their sexual behaviours – that is, hyposexuality or hyper-sexuality – after a brain injury or disease. Others can experience such changes as side effects of neurosurgery, which is where the next two chapters will take us.

3

SEXUAL SIDE EFFECTS
OF SEIZURE SURGERY

Many people, myself included, are in awe of neurosurgeons. They have been likened to gods, and I can see why. Your brain is the part of you that is *you*, and neurosurgeons touch people's brains every day. They suck and cut bits out of them, put things into them, relieve pressure from them. By doing this, they save lives. How could one not be in awe of those who deal with brains in such an intimate way?

One thing about neurosurgeons that I learned early in my career as a clinical neuropsychologist, however, is that many of them do not care much for my field. Post-operative changes in cognitive or thinking skills such as memory are just not relevant to them. Their job is to deal with life and death. If you survive your neurosurgical operation and are walking and talking after it, the neurosurgeon's job is done. One of my patients who experienced a dramatic change in her sense of self after a ruptured brain aneurysm told me that when she described this to her neurosurgeon, he responded, 'I deal with the nuts and bolts. Anything after that, I'm not your man.' Fair enough. Even gods have their areas of expertise.

I had a little run-in with a neurosurgeon at my first big overseas conference, nearly two decades ago. I was an excited and eager PhD student at an international epilepsy conference in the United States, proudly displaying my research findings on people who had experienced hyper-sexuality after temporal lobe surgery for epilepsy. He was a neurosurgeon who worked at an Australian hospital and I recognised him from local conferences. He marched up to the poster that displayed my research and ignored me while he read the title and conclusions. He smirked, then said aggressively, 'None of my patients have ever reported that.'

'Have you ever asked them?' I replied. He ignored me, turned away and walked off.

I can't remember if anyone else came to look at my poster after that because I was too busy fuming. Here I was on the other side of the world, and I'd been rudely snubbed by someone from my own country! I doubted he had ever asked his patients about any post-operative changes that were not strictly in his domain. He was only interested in two things: are they alive, and are they still having seizures? Sex was certainly not on his radar.

I recently met with a neurosurgeon and had the chance to ask him why he chose his profession, and whether any of his patients had ever discussed post-surgical changes in their sex drive. The conversation was much longer and more pleasant than the one I'd had 20 years before. John Chris-tie arranged to meet me one Saturday morning at a bakery, where we chatted over coffee and sourdough toast. He had just finished his weekly five-kilometre run, and I overheard him say to someone in the bakery queue that running was his 'new religion'. He was relaxed, softly spoken and gentle in his manner, apologising for wearing his sweaty sports gear as we sat down.

John had finished his neurosurgery training in the late 1980s, so had been working in the field for 30 years. I couldn't help but look at his hands and marvel at the fact that they had touched hundreds of brains. I asked him whether neurosurgery attracted people with a particular personality. He responded that neurosurgeons were typically 'driven and obsessive', particularly in the time before brain imaging revolutionised their specialty. It was 'intense and solitary work', he said, as there were usually only one or two neurosurgeons per hospital. His boss had been the only neurosurgeon in his regional area of 400 000 people for 12 years. Needless to say, as soon as John arrived after completing his training interstate, his boss immediately went on holiday!

John acknowledged that neurosurgeons are not generally renowned for their communication skills. They don't have time to chat with patients about the things that occupy a neuropsychologist's time; John usually has 15 people waiting to be seen in his neurosurgery ward, and as he does his rounds he moves 'from one devastating situation to another'. 'I might have a young child with a tumour and their distraught family in one bed, which of course really affects me, but a minute later I need to be composed and able to talk to my next patient,' he explained. That's why he sticks to the nuts and bolts.

We discussed how revolutionary neuroimaging – brain scanning technology – has been for his specialty. Before brain scans, neurosurgeons had to rely on clinical history and examination of their patients, and it was often difficult to diagnose and localise brain lesions and masses. If you were lucky you might have been able to get an angiogram, a type of X-ray that shows the location of any abnormal blood vessels, to give you a clue about where a brain mass like a tumour might be. If a patient had sustained a traumatic brain injury and was unconscious, it could take 'all night to sort out where the bleed was – with exploratory burr holes all over their skull', John recalled. (Burr holes were holes drilled in the skull to check where a bleed in the brain had occurred.) 'Now you can just run people through a scanner.' Something that took hours of blind exploration is now solved in the time it takes to complete the scan.

When it comes to his patients' sex lives post-surgery, John said it was 'not a question I would ask and something I think they would be hesitant to bring up'. He could recall discussing this with only one patient in three decades. The husband of a woman who had had a brain haemorrhage reported that she had 'no interest in sex' a few months after surgery, when previously she had been 'very keen on it'. John saw her in a social setting a couple of years later, and she had her 'vivacious personality' back, but he never followed up about her sex life. He only ever brings up sex in a consultation if the patient's operation involves the pituitary gland, as this can cause hormonal changes that can impact on sexual function. On the rare occasions when he does ask, he said, he only wants 'to know if it works or not'.

John described his most memorable patients as 'heroes', the 'amazing ones who carry on despite their awful situations'. His favourite thing about his job, and the thing that attracted him to it in the first place, is 'microsurgery – seeing beautiful brain anatomy under the microscope and removing a tumour that is challenging to reach'. What he doesn't like doing is operations that have a higher risk–benefit ratio, where the risks are high and the potential benefits small, such as some patients with degenerative spinal conditions 'who want surgery, but surgery won't help them', he said. 'It's always tricky when someone pays you for something and you can't really do anything for them.'

When I had first contacted John to ask him if I could interview him, he said he was 'not a typical neurosurgeon'. When I asked him what he meant by this, he couldn't recall saying it, but suggested that maybe he wasn't as driven as some of the others. On my way home after the interview, I knew why. Unlike 'typical' neurosurgeons, he actually was a great communicator. If anyone in my family ever needed neurosurgery, I knew who I would ask for.

FOR MOST PEOPLE, THE WORDS 'SEIZURE' AND 'EPILEPSY' trigger visions of someone dramatically falling unconscious to the ground and shaking violently, with saliva foaming at their lips. This is only one type of seizure, known as a 'generalised' seizure. Usually the neurons in our brains are firing at different times and in different parts of the brain, like random fireworks, but during a seizure the neurons' firing becomes synchronised – they all fire at once and together. It is this sudden synchrony that results in a seizure. When this occurs across the whole brain it is called a generalised seizure, which dramatically affects the whole body.

But there are other types of seizures where the synchronous electrical activity remains 'focal', or only in one part of the brain, so the effects are subtler. For example, temporal lobe seizures are restricted to the temporal lobe, and they cause less obvious changes in behaviour. People who have these seizures describe feeling like they have 'missed time'. Observers say that they stare blankly or appear vague. They might pick at their clothing. They often look like they are chewing something. These seizures typically last a few minutes. They are unpredictable and can occur at any time of the day or night.

After a temporal lobe seizure, the person can feel dazed, confused and tired. If the seizures occur multiple times a day, life can be very difficult. People who experience these seizures cannot drive and are restricted from many types of employment. So it's not surprising that anxiety and depression are common in people with temporal lobe epilepsy. As for their sex lives, they are typically hyposexual. There are likely to be many causes of this hyposexuality, including dysfunction of the temporal lobe and structures within it, particularly the amygdala – all crucial parts of the sexual neural network (see Chapter 1); disrupted sex hormones due to seizure activity; and mood symptoms and psychological changes, such as reduced self-esteem and confidence. Would you feel sexy if you knew you might have a seizure at any moment?

You only get to be considered for a 'temporal lobectomy', or removal of the temporal lobe where the seizures are coming from, if your seizures are 'intractable' – that is, severe, frequent and unable to be controlled by anti-seizure medication. There is no doubt that one of the most frightening things you could ever experience in life would be to have a piece of your brain removed. But if it meant a chance at a cure, a chance at a seizure-free life, it would be a risk worth taking. To be considered for surgery you have to submit your brain for investigation by an epilepsy surgery team. This is a group of brain specialists who are all searching for the same thing: the place in your brain where the seizures start, or the 'seizure focus'. If they find it, the team then needs to assess if it is safe to remove this bit of brain. Neurologists, neuroradiologists, neuropsychologists – all these 'neuro' doctors – are like a group of detectives on the hunt for a murderer, but their target is the seizure focus. They run every known type of neuro investigation to study the function and structure of your brain.

If you are being considered for surgery, you typically will spend at least a week in hospital beforehand, wired up to an electroencephalogram (EEG) machine that monitors electrical activity in your brain. You go off your anti-seizure medications; you need to have as many seizures as possible so the EEG can record what is happening in your brain during your seizures, allowing the focus of the seizures to be detected. Your brain is carefully scanned by all the methods known to humankind, assessing blood flow, glucose metabolism, electrical activity, morphology, metabolites and volume. Every centimetre of your brain is measured and viewed by brain imaging experts, all searching for that 'brain murderer' – the seizure focus.

The culprit? Typically it is hippocampal sclerosis, or a hardening of the hippocampus, that small structure in the temporal lobe snuggled up near the amygdala. A sclerotic hippocampus can play havoc with the electrical activity in this brain region, causing temporal lobe seizures. If anti-seizure medications don't work, a unilateral (one-sided) temporal lobectomy, where the affected lobe is removed, may be the only remaining chance to stop the seizures. This surgery is known to have been done in antiquity; even prehistoric human remains from the Neolithic period have been found with evidence of a procedure called trepanation, where holes were drilled, bored or scraped into the skull, just like the burr holes that John Christie referred to. Ancient cave paintings suggest that people believed that this could cure epilepsy, migraines and mental disorders. Neuro-surgeons have of course adapted their techniques over time, assisted enormously by the technological advances in neuroimaging; nowadays the scar tissue or the sclerotic hippocampus itself is typically sucked out.

Surgeons have learned the hard way not to remove both temporal lobes. Henry Molaison – 'HM' – had surgery to both his temporal lobes in an attempt to cure his severe seizures, but the surgery left him with a complete inability to form new memories (see Chapter 1). He was studied extensively and contributed to much of our knowledge about the crucial role that the temporal lobes play in our ability to remember. The risk to memory function is now carefully considered in the medical work that occurs in the lead-up to a temporal lobectomy. Those considering surgery receive counselling about what to expect in regard to their memory and other thinking skills. In some surgery programs, this pre-surgical counselling extends to other domains, covering potential changes to mood, psychological and social functions, and includes ongoing follow-up after surgery.

Surgery for temporal lobe epilepsy can lead to a range of changes in a person's life as well as their brain. They will hopefully be seizure-free after the surgery, but there can be some risks to certain cognitive functions, such as language (specifically finding words), and memory (in particular verbal memory or the ability to learn and recall verbal information) if the left temporal lobe is removed. This is why clinical neuropsychologists are part of the 'neuro' team involved in the pre-surgery evaluation, as they need to check the person's cognitive functioning and see what the risks of surgery will be to their thinking skills. There are also behavioural and psychological effects that can occur after this type of surgery. My thesis supervisors discovered and characterised a post-surgical phenomenon called the 'burden of normality', a syndrome that can arise from the process of adjusting after effective seizure surgery. A person whose surgery has been successful may describe feeling like a 'new' person, with increased self-confidence and improved self-image, while others may mourn the loss of their epilepsy and have difficulty discarding the 'sick' role. A sense of grief for the years lost due to the limitations of epilepsy may result in the need to 'make up for lost time', which can manifest in an elevated mood, or anxiety and depression, or in behaviours like excessive working, exercising, socialising or domestic work. Others may feel overwhelmed with the new demands of post-operative life and engage in avoiding behaviours. Family dynamics often change; as the seizure-free person exhibits increased independence, others who have acted as carers may also need to redefine their own roles.

These clinical features of the burden of normality are commonly reported after temporal lobectomy and affect up to two-thirds of patients, and can continue for years after the surgery. Changes in sex drive after surgery can be considered as part of the process of adjustment, and be conceptualised as a behavioural feature of the burden of normality. The changes that people have reported in their sex lives include dramatic increases in sex drive, constant sexual thoughts and fantasies, and (more rarely) shifts in sexual orientation.

One of the most dramatic sexual outcomes after temporal lobe surgery was in Sharon, a 31-year-old woman who'd had measles encephalitis just before she was three years old, and developed seizures a year later. They typically occurred a few times a week. At the time of her operation, a left temporal resection, Sharon was in a longstanding cohabiting but platonic relationship with an older man, who she described as overprotective and managerial; she revealed that she planned to leave him after her surgery. Her seizures stopped after surgery. Three months after the operation she described feeling like a new person, with a greater sense of self-confidence and independence. She now viewed her partner as a 'father figure' and eventually ended their relationship. Nine months after her surgery she reported hypersexuality expressed through frequent sexual thoughts and fantasies, in addition to a change in sexual orientation from heterosexuality to homosexuality. This was described as a longstanding but latent orientation which could finally be revealed, she said, as the 'new me'. A year after her surgery she began to socialise with members of her local lesbian community and adopted a new self-image and lifestyle. For the three years that she was followed up after her surgery, she was seizure-free.

In another case, Peter, a married 29-year-old, underwent a right temporal resection after having seizures since he was two years old. He was seen for over four years after his surgery and in that time remained seizure-free. Three months after his operation he described a new outlook on life; he had the sense that he was facing a 'whole new world', and felt he had been given a 'second chance'. He reported an increased desire for sex with his wife, and sexual attraction to female work colleagues. A year after his surgery he resumed a previous homosexual relationship. He reported frustration that his wife had not had the same operation to increase her sex drive too. Four years after his operation he had an intense heterosexual affair and stated that he was leaving his wife. Peter occasionally lamented the loss of his simple, monogamous pre-operative life, when he wasn't consumed with sexual thoughts.

FOR THREE YEARS, I INTERVIEWED PEOPLE WHO HAD undergone epilepsy neurosurgery about their sex lives. These interviews took place well over a decade ago, but some of the patients are still clear in my mind. People ask how I managed to talk to people about something so personal, and I can't pretend I wasn't nervous the first time. But my nerves dissipated when I realised how grateful and relieved these patients were to discuss it. I met with 74 people who had undergone epilepsy surgery, and all of them were willing to talk about sex. More than half of them had experienced a post-operative change in their sex drive. Some patients expressed a wish for their partners to have the same operation to 'keep up with them' sexually, just as Peter did. But not all of them had dramatic hypersexual changes; most experienced a subtler increase from their pre-operative hyposexuality to what they considered to be a normal frequency of sexual activity. Sexual changes were more common in people who had temporal lobe surgery than in those who had surgery to other parts of the brain. This was what we had predicted, given that we already knew that the temporal lobe is crucial to the sexual neural network (see Chapter 1).

Sexual changes were more commonly reported if the surgery was to the right temporal lobe rather than the left. This supported earlier research on the role of the right side of the brain in emotional processing: sex involves complex emotions, so it is no surprise that there may be right-sided dominance in mediating sexual function, making sexual changes more likely when there is a dysfunction in the right temporal region. Women are more likely to report hyper-sexual changes after neurosurgery for epilepsy, and seizures that cause sexual arousal or orgasm (as discussed in the previous chapter) are also more common in women. This may relate to differences in brain morphology between men and women; one study in the 1980s found that seizures causing sexual arousal were reported almost exclusively by women, and proposed that these findings 'suggest that the neural organisation of psychosexual behaviour differs in human male and female brains'.

But does it? There's no simple answer yet; most neuro-imaging studies of human sexual behaviour focus on only men or women, which makes it difficult to compare their results. Those studies that have compared how the brains of men and women activate in response to visual sexual stimuli have found many brain regions display similar activation in men and women, regardless of their sexual preference. The question of whether there are gender-based differences in brain function and structure is still hotly debated, and is another topic that warrants a whole book of its own. A recent book by neuroimaging professor Gina Rippon, *The Gendered Brain*, argues that the research into the differences between male and female brains is nothing short of a mess – full of misinterpretation, publication bias, weak statistical power and inadequate controls.

In the case of sexual seizures and my own research on hypersexuality after seizure surgery, it certainly seems that these are more likely in women, but how this relates to brain differences in men and women is unclear. Curing people of their seizures is the main purpose of this type of neurosurgery, regardless of gender, and seizure freedom is the outcome that neurosurgeons are focused on. But this is not the only outcome relevant to people. Their sex lives matter too, and they need to know that their surgery could change the way they behave sexually. Typically these changes are for the better, but patients need to be prepared for the possibility of more dramatic changes too. As John Christie said, this may not be a neurosurgeon's domain, but if their patients are referred to an appropriate specialist – such as a clinical neuropsychologist! – they will at least be able to discuss these issues and obtain valuable information for their post-operative lives. I was lucky to be exposed to a world-renowned epilepsy surgery team during my PhD; they recognised the importance of this for patients. Of course I am biased in my views, but I cannot see how the outcome for patients can be optimal without such care.

Seizure surgery is not the only type of neurosurgery that can result in sexual changes; treatments such as deep brain stimulation and the insertion of shunts in the brain can also trigger hypersexuality. Horrifyingly, brain surgery has also been performed with a specific sexual outcome in mind – to 'treat' or change sexual orientation. These neuro-surgical interventions are explored next.

4

STIMULATION, SHUNTS AND PSYCHOSURGERY

Ray still dreams of dancing. In his twenties he was a ballroom dancing champion. That's how he met his wife, Rosa. They were paired up at the local dance hall as shy teenagers and learned all the moves together. It wasn't long before their partnership became more than just a dancing one, and as their love bloomed they were thrilled to win local and then national competitions. Ray can still remember the feel of it – the gliding, twirling, lifting, twisting. He didn't have to think about moving; his feet and legs just did it. The dance instructor called it 'flow'. The competition judge also used that word when they won their biggest trophy, saying they were 'a pair in perfect flow'.

That flow had long gone from Ray's body, replaced with other f-words: freezing, festination, falls. He could think of one more f-word to describe his life, and frequently used it when trying to move. Just the short walk from his bed to the toilet was a struggle. 'Fuck! Why won't my bloody legs work?' he'd cry out in frustration. Rosa would sigh, rush to his side, put her arm around him and ask if he felt like a cup of tea. He remembered when he'd first heard the neurologist say the words 'Parkinson's disease' and how he hadn't heard anything more after that. He remembered Rosa's tears dripping off her chin, and that he'd held onto her hand a bit tighter than he usually did.

The idea of brain surgery scared him. 'Deep brain stimulation' sounded like something from a science fiction movie, but the thought of never feeling 'flow' again scared him more so he was willing to try it. The idea of someone snooping around deep within his brain was weird. But the motor symptoms of his Parkinson's disease were severe, and he had not had much relief from the cocktail of medications he had tried, so his neurologist had told him it was worth a try. 'Who knows,' he asked himself, 'maybe Rosa and I will actually get to dance together again?'

On the day of the treatment he kissed Rosa goodbye and told her he looked forward to dancing with her soon. She wiped a tear away as he shuffled through the automatic doors.

And dance he did. The treatment seemed miraculous. It was as though he had a new body. His shuffling gait – the much-feared festination – was replaced by smooth strides. He couldn't stop smiling. He didn't walk; he preferred to dance. Rosa indulged his constant grabbing of her waist and insistence on waltzing around with her. She was happy that he was happy.

But this miraculous treatment had a hidden side effect that they were both struggling with. Not only was Ray insisting that Rosa dance with him all the time, he was also insisting on sex. Multiple times a day. It was fun at first, but after a month Rosa started to get worried and wondered if it would ever settle down. She was exhausted. When she said no a few times in a row, she caught him in the garage masturbating over an old porn magazine. She suspected that if she didn't oblige with sex, he might seek it elsewhere. She was right. When she saw a series of unusual transactions in their bank account, she became suspicious and rang the bank, then traced down the name on the bank statement to a local brothel. She stopped making him cups of tea after that.

One night she heard Ray come to bed and kept reading her book. He lay with his face down in his pillow, and she felt the bed wobble. 'I'm so sorry,' he said between sobs. 'I can't help it. There's something wrong with me. We have to tell the doctor. I'm scared they won't let me have the treatment anymore, but I can't go on like this.' She put her book down and nodded, and then buried her head in her own pillow to soak up her tears.

Hypersexuality is the most common of the impulse control disorders (ICDs) that can occur in Parkinson's disease, either in response to dopamine medications or after deep brain stimulation treatment. It has been reported that around 3 per cent of people with Parkinson's who are taking dopamine replacement therapy will experience hypersexuality at some point; it is more common in men, and in those taking higher doses of dopamine agonists – that is, drugs that mimic dopamine. Other common ICDs are pathological gambling and compulsive shopping; some studies report that up to 40 per cent of people with Parkinson's disease are affected by ICDs of some kind.

The catch with deep brain stimulation as a treatment is that while it is highly effective at treating motor symptoms, and can also be prescribed to treat ICDs, in some cases it actually makes them worse – or triggers new ICDs. According to studies of this phenomenon, ICDs are more common in males who have a younger onset of Parkinson's disease and have a previous psychiatric history or family history of addiction. Why some people with Parkinson's develop new ICDs after surgery is not known, but all the ICDs are a result of tampering with the structures known to be part of the reward-processing system in the brain. In the case of hypersexuality, it is thought to reflect the direct stimulation of the structures within the sexual neural network.

Deep brain stimulation involves the insertion of electrodes within certain brain areas to produce electrical impulses. The electrodes are controlled by a device like a pacemaker that sits under the skin of the chest and is attached to the electrodes by a wire. Deep brain stimulation for Parkinson's disease typically targets the subthalamic nucleus, which is part of the basal ganglia, a complex structure deep within the brain. The subthalamic nucleus forms part of the frontostriato-thalamic-cortical loops, which mediate motor, cognitive and emotional functions, so it is not surprising that stimulating it can result in behavioural changes. Deep brain stimulation can also cause increased sensitisation of the brain to dopamine therapy, which is usually still required after the stimulation, although typically at a lower dose than that needed pre-surgery.

It is often partners who report ICDs rather than the patients themselves; the impact of these behaviours can be highly distressing for family members. Managing ICDs in people with Parkinson's disease is a juggling act. It involves weighing up the severity of their motor symptoms, how they respond to medications and the impact of the ICD on their lives. For those with Parkinson's disease contemplating deep brain stimulation, the old phrase 'knowledge is power' is applicable. Being informed about the possibility of hypersexuality and other ICDs can reduce the shock if these side effects occur.

APART FROM SUCKING OUT AND STIMULATING BITS OF brain, neurosurgeons also need to put things into brains. One example is called a shunt, which is used to treat hydrocephalus – literally, water (*hydro*) on the brain (*cephalus*). Our brains are bathed in a liquid called cerebrospinal fluid; the brain stays moist thanks to a finely balanced system of production, absorption and flow of this fluid. Special chambers within our brains, the ventricles, are like pools. Typically the cerebrospinal fluid fills and flushes through these ventricles in a constant tide-like flow, but sometimes the flow is disrupted due to a blockage, resulting in hydrocephalus. The cerebrospinal fluid build-up can cause the ventricles to enlarge and put pressure on the brain, resulting in symptoms such as visual disturbance, walking difficulties and incontinence. Treatment may be required and involves a neurosurgeon inserting a small tube – a shunt – that helps re-establish the flow of cerebrospinal fluid.

A change in sex drive is not something you would expect as a result of having a shunt inserted by a neurosurgeon, but this is exactly what happened to two elderly gentlemen, Albert and Arnold. They were both in their seventies. Albert had never been married and it was noted that he had 'courted very little' and 'never used coarse or suggestive language'. After he collapsed in the nursing home where he lived and was found to have hydrocephalus, he had a shunt inserted – and suddenly developed a sexual appetite. He approached and fondled female patients, crawled into their beds with sexual intent and used sexually explicit language. Albert had to be restrained continuously to prevent his constant sexual demands.

Arnold was married and, according to the case report, he and his wife had 'weekly sexual relations'. He developed herpes encephalitis and after waking from a coma, he began making sexual comments to women in the hospital, fondled the nurses and masturbated in public. Given that we know herpes encephalitis affects the temporal lobes (see Chapter 1), it is likely that his initial manifestation of hypersexuality was due to temporal lobe damage. After treatment of the acute phase of his encephalitis he developed ventricular enlargement and needed a shunt to be inserted. Following this procedure, his sexual disinhibition was exacerbated; his wife said he was 'disgusting' and became 'the man with a thousand hands'. He tried to fondle her every time she was within reach, demanded sex multiple times a day, and asked if she could have sex with other men while he watched, an interest he'd never expressed before in their decades of marriage. This excessive sexual interest had been present for two years at the time he was seen by the doctors who wrote up his case study. No doubt it was an exhausting time for his wife.

The common thing in both Alfred's and Arnold's cases was that the tip of their shunts ended up in the septum. This is a subcortical structure comprised of two parts: the septum pellucidum and septum verum. The septum pellucidum, a thin and almost transparent membrane that runs down the middle of the brain, is surrounded by a collection of nuclei (the septal nuclei), which are extensively connected with other parts of the sexual neural network, including the amygdala, hypothalamus and the ventral tegmental area (which, along with the substantia nigra, is an area of the brain that releases dopamine). Once again, we see that these brain structures are all part of a widespread interconnected network; a shunt tip disturbing the septal region or any other part of the sexual neural network could lead to unusual sexual side effects.

THIS WAS NOT THE FIRST TIME THAT TINKERING WITH the septal region had been found to result in sexual disturbances. Other observations had already proved that this brain region was part of the sexual neural network. In the 1950s and '60s the controversial American psychiatrist

RG Heath electrically and chemically stimulated different brain regions in 54 patients with various conditions, including schizophrenia and narcolepsy (a disorder characterised by excessive daytime sleepiness, sudden 'sleep attacks', and in some cases a loss of muscle control called cataplexy). All of Heath's subjects showed what he called a 'pleasurable response', with varying degrees of sexual arousal, when their septal region was stimulated. Heath developed a self-stimulating device that was attached to intracranial electrodes – that is, electrodes that were placed within the brain, usually to find the origin of seizures in people with epilepsy. This enabled patients to press a button to self-stimulate parts of their own brains. When the electrodes were in the septal region, patients could give themselves an immediate orgasm. Not surprisingly, some patients pressed this button repeatedly, finding the instant pleasure impossible to resist.

The term 'psychosurgery' brings to mind tragic cases of frontal lobotomy, such as that portrayed in the Oscar-winning film *One Flew Over the Cuckoo's Nest* (1975) and the 1962 novel it was based on. Jack Nicholson's character, Randle Patrick 'Mac' McMurphy, has faked mental illness to be placed in a psychiatric hospital rather than jail, but ends up being forced to have a frontal lobotomy, or leucotomy, a procedure that severs the connections between the frontal lobes and the rest of the brain. The result is that McMurphy is distressingly reduced to a vegetative state, no longer the charismatic criminal who fearlessly stands up to authority.

In the early 1970s, Heath used his program of septal stimulation in a bizarre and shocking way: in an attempt to 'initiate heterosexual arousal and behaviour in a homosexual male'. This mirrored the disturbing psychosurgical techniques used in Germany to 'treat' homosexual men in the 1960s and '70s. Men who were considered to have 'deviant' sexual behaviours were subjected to a brain surgery technique called stereotaxic hypothalotomy, which involved the surgical destruction of a part of the hypothalamus called the ventromedial nucleus. The idea was based on earlier experimental animal research on cats whose amygdalae had been removed and had exhibited hypersexuality; removing their hypothalamus had been found to reverse the hypersexuality. The surgeons who conducted hypothalotomies on humans considered the hypothalamus to be the 'sex behaviour centre', and they reported that its destruction led to abolished or weakened sex drive.

The hypothalamus is a cone-shaped structure deep within the brain which is involved in the control of crucial body functions including the autonomic nervous system and endocrine functions (see Chapter 1). Most of the 70 or so men (and one woman, according to some reports) who had stereotaxic hypothalotomy surgery were homosexual and were in prison or another institution. Some had engaged in paedophilic behaviour; others had not. In the 1960s and '70s, homosexuality was still widely considered to be something that needed 'treatment'. It was even listed as a disorder in the *Diagnostic and Statistical Manual of Mental Disorders* (DSM), the psychiatry bible first published in

1952, until 1973, and remarkably wasn't removed from the World Health Organization's International Classification of Diseases (ICD) until the tenth version published in 1992. It is clear that homosexuality was considered 'abnormal' by the surgeons who performed hypothalotomies.

This 'sexual psychosurgery' was highly controversial. A guest editorial in the journal *Archives of Sexual Behavior* in 1979 argued that the surgery's theoretical basis was highly questionable as it demonstrated an 'inadmissible transference of animal-experimental findings to human beings', and ignored psychological and sociological knowledge about human sexual behaviour. The scientific standard of the reports about the surgery were heavily criticised. The post-operative examinations done by the surgeons themselves appeared 'unsystematic, coincidental and random'. Reports on the outcomes of hypothalotomy were scant and biased in that they only addressed positive outcomes and mixed quantitative changes in sex drive with qualitative changes in sexual orientation. There is no information about the types or extent of psychological changes. However, the unwanted or negative effects of the surgery, rarely mentioned, included a total loss of sex drive, dizzy spells, ravenous hunger, weight gain and increased verbal aggression. One case refers to a person's loss of ability to remember dreams. Numerous ethical concerns were raised. While the surgeons claimed that their patients were suffering and were operated on 'at their own initiative', the authors of the journal editorial, Inge Rieber and Volkmar Sigusch, maintained that the patients could not voluntarily decide on the surgery, as most were prisoners sentenced to punishment and they 'hoped that, through the surgery, they could gain their freedom'. Rieber and Sigusch argued that 'this kind of surgery under these conditions is virtually an act of coercion', and released a public statement calling for it to be ceased immediately. Public outcry eventually led to the end of this type of psychosurgery.

The German patients who were given hypothalotomies 50 years ago are the only people known to have undergone neurosurgery for the sole reason of specifically targeting their sexual behaviour. Thankfully, in many countries, the destructive notion of homosexuality as a 'disease' requiring 'treatment' has been abandoned, but it is frightening that homosexuality remains illegal in around 70 countries, and even more horrifying that it is punishable by death in at least ten countries. The mistaken and deeply damaging belief that homosexuality is something that needs to be 'cured' still exists, and can be seen in the various types of 'conversion therapy' that are practised all around the world. These alarming therapies range from 'spiritual' interventions to physical measures such as electric shock therapy.

Shock therapy to intervene in sexual orientation has been portrayed in *Masters of Sex*, the TV series based on the work of researchers William Masters and Virginia Johnson in the 1950s and '60s (see Chapter 5), and there are reports of people still being subjected to this today. The film *Boy Erased* (2018), based on a memoir by Garrard Conley, recounts Conley's traumatic experience of attending a gay conversion therapy program. Conversion therapies have been banned in many countries but remain a disturbingly widespread practice, and one that has led many to suicide. They have been labelled 'torture' by the United Nations, and there are calls for such programs to be outlawed worldwide. Let's hope the future is more promising than that portrayed in the TV series *The Handmaid's Tale*, in which a lesbian character is charged with 'gender treachery' and sentenced to 'redemption'. She watches her lover being hanged and suffers the 'punishment' of genital mutilation.

FOR PEOPLE WITH NEUROLOGICAL CONDITIONS SUCH as Parkinson's disease, like Ray, or temporal lobe epilepsy like Sharon and Peter (see Chapter 3), neurosurgery may be the final attempt to remove or control symptoms of their brain disorder. For others like Albert and Arnold, with a life-threatening condition such as hydrocephalus, neurosurgery is required for their survival. Dramatic sexual side effects like those I've described are usually rare and unexpected. Whether they are considered unwanted or desirable effects will no doubt depend on the nature of the sexual change and, if the person has a partner, how the partner feels about the change. Brain surgery is a life-changing experience for the person undergoing it, but it also impacts their partners too.

When one person in a couple is about to undergo brain surgery, sex is probably not high on that couple's list of priorities. After the initial recovery phase, though, they might wonder when it's safe to 'get back in the sack'. This question was posed in a recent letter to the journal *Neurosurgery* by two UK-based specialists. The authors stated that although many neurosurgical patients are 'young, fit, and lead active sexual lifestyles', there is a lack of evidence that addresses the issue of when or if sex is safe after neurosurgery. They refer to the 'problem of pressures', particularly the rise in intracranial pressure – or pressure inside the skull, due to increase in volume of brain tissue, cerebrospinal fluid or blood – that occurs during sex and orgasm. Higher intracranial pressure reduces cerebral perfusion, or oxygen supply to the brain, which increases the risk of stroke or seizure in a post-operative brain that is sensitive to oxygen levels. Yet there can also be pain-relief benefits from the endorphin response triggered by orgasm. The authors concluded that neurosurgical teams need to be prepared to discuss sex with their patients 'without any stigma attached', that a patient's fear of sex after neurosurgery should not be dismissed, and that patients should be encouraged to resume their sex lives 'when the patient is up to it, with [a] common sense approach'.

The cases I've described demonstrate how neurosurgical stimulation of certain parts of the sexual neural network can alter our most intimate sexual thoughts and behaviours. But can it work in reverse? Can sex alter our brains? Yes, it can – and in very dramatic ways. The next chapter will explain.

5

CAN SEX CHANGE YOUR BRAIN?

In 400 BC, the Greek physician Hippocrates wrote a treatise on epilepsy called *On the Sacred Disease*. At the time, epilepsy – a word from the Greek meaning 'to seize, attack, take hold of' – was widely believed to be caused by attacks from demons or gods. Hippocrates challenged this and proposed that the brain itself was the cause of the disorder, not invisible deities. Five hundred years later, Galen – now considered the greatest physician of the Roman Empire – agreed.

Yet Hippocrates and Galen were well ahead of their time in proposing that epilepsy was caused by physical abnormalities originating in the brain. The majority of physicians and the general public believed that it was caused by supernatural interventions, sexual activity such as masturbation, and these beliefs persisted well into the nineteenth century. In fact, of all neurological diseases, epilepsy is the one that has been most frequently linked to sex. One influential medical voice on this topic was the famous

Swiss physician Samuel-Auguste Tissot, who argued that excessive masturbation could cause epilepsy. He published a book in French in 1760 called *L'Onanisme*, translated into English as *Onanism: Or a treatise upon the disorders produced by masturbation*. (I was amazed to find you can still buy a paperback copy through Amazon.) At the time, castration and clitoridectomy (removal of the clitoris) were reportedly performed on people with severe epilepsy.

The supposed link between epilepsy and sex continued into the next century when neurologist Edward Sieveking emphasised sexual disturbances as a leading cause. In his book *On Epilepsy and Epileptiform Seizures*, published in 1858, he stated that although he believed hereditary factors played a role in the disease, 'the unanimous consent of all writers on epilepsy demonstrates the truth of the statement that in this disease, the sexual organs are very frequently at fault'. He cited an ancient proverb attributed to Galen, '*Coitus brevis epilepsia est*' – 'Sex is a brief seizure', and argued that sexual derangement 'enfeebled the system, and by producing excitability gives rise to the epileptic paroxysm'. Sieveking regarded masturbation as a specific cause of seizures, and wrote that in nine of the 52 people he had treated for epilepsy, 'the sexual system was in a state of great excitement, owing to recent or former masturbation'.

I was surprised to learn that Sieveking had been a physician at the National Hospital for Neurology and Neurosurgery in London in 1864. I worked at the same place over 150 years later. It was where I'd assessed Jack, the man exhibiting hypersexuality after a fall from scaffolding, described in Chapter 2. I knew that many renowned neurologists had worked there, including John Hughlings Jackson (1835–1911), often referred to as 'the father of English neurology and modern epileptology', and William Gowers (1845–1915). Neither of these influential neurologists considered sex the origin of epilepsy. Rather, they identified neurophysiological causes and laid the foundations for current views of the aetiology of epilepsy.

Yet while the notion that sex causes epilepsy has been well and truly debunked, it is widely accepted and scientifically proven that sex can actually trigger some brain conditions. These dramatic brain disorders can be temporary, as in the case of transient global amnesia, or can cause permanent brain damage, such as the rupture of a brain aneurysm. In the rare cases where these conditions are sex-induced, it seems it is the dramatic blood flow changes in the brain that occur during sex that are to blame.

'WHERE AM I?' SARAH MUMBLED.

Joe laughed. Sarah had a quirky sense of humour and he thought she was joking. He had his eyes closed and had been enjoying a post-sex snooze. He ran his fingers along her thigh and replied, 'You're a funny one.'

'Where am I? Why am I here?' she asked, louder this time and more insistently.

Joe laughed again. 'What are you playing at, you cheeky thing?'

It wasn't until she asked a third time, and he heard the tremor in her voice, that he realised it wasn't a joke. She was genuinely frightened. 'What's happened? Where am I?'

Joe rolled on his side to face her. Sarah was staring at the ceiling. Tears were trickling down the side of her cheek, forming a wet patch on the pillow beside her face.

'Sarah, you're in bed. What's wrong? Why do you keep saying that? What's going on?'

'Why am I here?'

They had just had sex – very successful sex, Joe felt, given that she'd had an orgasm for the first time in months. He felt proud that he still had the knack to get her there. But his post-sex pride was quickly evaporating and being replaced by worry. He put his hands on her cheeks and touched his nose to hers. Then he pulled back and stared into her eyes.

'Darling, you're at home in bed with me. Are you feeling OK?'

Sarah looked at him blankly, and when she said exactly the same words again, he became frightened. Was she having some kind of stroke? He tried to think back to his first aid training at work. What was the acronym you had to run through for stroke? Something about asymmetry and paralysis? He checked her face again and she looked fine, apart from the confusion in her eyes. He told her to move her arms and legs and asked if they felt normal.

'They feel fine,' she replied, 'but where am I?'

Joe was perplexed. 'Why do you keep asking me that?'

'Why am I here?'

He felt sick. 'Just wait here, darling. I'll be back in a minute.' He grabbed his phone off the bedside table, pulled his dressing gown on and rang an ambulance.

Transient global amnesia is a sudden and temporary impairment of memory during which a person cannot learn or recall new information (anterograde amnesia) and may also have difficulty recalling old memories (retrograde amnesia). During transient global amnesia the person remains alert and aware of their personal identity and is able to communicate. The main symptoms are repetitive questioning and disorientation. It typically occurs as a single episode and is considered a 'benign' disorder, as the memory problems completely resolve within 24 hours and there are no long-term effects. In the case of Sarah, she was back to her usual self the next day, and eventually her and Joe were able to laugh about the time their sex was so good she 'lost her mind'.

Many people have learned about amnesia from characters in films such as *50 First Dates* (2004) and *Eternal Sunshine of the Spotless Mind* (2004). Although entertaining, 'cinematic amnesia' is usually not realistic. In an article in *The British Medical Journal*, neuropsychologist Sallie Blaxendale summarised 'amnesia in the movies', and pointed out the many myths that are perpetuated on screen. For example, characters with amnesia have often suffered a head injury that has caused them to forget who they are (loss of identity) and lose memories of their whole life before their injury (retrograde amnesia), but they have no difficulty learning and recalling new things after their injury. In reality, extensive retrograde amnesia and loss of identity are rare and typically suggest a psychological cause rather than a brain injury. The more typical type of amnesia after a brain injury involves difficulty learning and recalling new things (anterograde amnesia). It is also common in movies for a second head injury to cure the character of their memory loss, which never occurs in reality.

An accurate cinematic portrayal of amnesia, however, is found in the film *Memento* (2000), in which the character Leonard has a very realistic amnesia after a head injury. He retains his identity but has significant anterograde amnesia, to the degree that he tattoos crucial new information on his body to aid his memory. In fact, the animated fish Dory from the children's films *Finding Nemo* (2003) and *Finding Dory* (2016) has another of the most neuropsychologically accurate portrayals of amnesia ever seen on screen. I don't know of any films that feature a character who experiences an accurate transient global amnesia, but there would be no need to exaggerate. Even a realistic portrayal of this bizarre condition would be riveting.

Although there have been hundreds of reported cases of transient global amnesia in the medical literature, its cause is still unknown, and it is regarded as one of the most mysterious of all the neurological conditions. A common risk factor is a history of migraines. Various triggers have been reported, including strenuous physical activity, emotionally arousing or stressful events, a sudden change in body temperature, and – yes – sex.

In a study that specifically explored what triggered transient global amnesia, sex was the most common precipitant in the 21 cases studied, reported by one-third of the subjects. Interestingly, this same study distinguished between sex that was 'conjugal' (within a married or established couple) or 'extra conjugal' (outside of a marriage or relationship – in other words, during an affair), and noted that being unfaithful could be considered a 'stressful situation'; it had been reported in several transient global amnesia cases. Sex ticks two of the boxes of the known triggers for this type of amnesia: it can be both physically strenuous and emotionally arousing. This is particularly so if the sex is with someone who is not your usual partner, as in the case of Roger.

Roger was 46 years old, a successful, hard-working architect and teacher. He was prone to self-doubt and anxiety, and had experienced panic attacks in his twenties. He felt stressed and overworked, and when his boss put him in charge of a large architectural project, he complained to his wife that he wished he could retire. He also started to exhibit more aggressive, erratic behaviour. When someone backed into his car in a parking area he exploded, gesturing wildly and screaming when the other driver approached him to apologise. One day he argued with the students in his architecture class, stormed out of the classroom and drove home to have lunch with his wife. After lunch, he told her he was going back to the office but he lied. He was actually going to visit his lover, who he had been secretly seeing for over a year. He felt he deserved some joy in his life, and it had been weeks since he had seen her. As he drove into her driveway, his stress slowly evaporated and was replaced with excitement. He felt like he was 17 years old again. They had frantic sex, like teenagers who were afraid of getting caught by their parents.

Afterwards, Roger couldn't recognise his own pants that were draped across a chair, and struggled to recall what he had done that morning. He eventually got his clothes on and was able to drive home. His wife noticed immediately that something was unusual about him. He was vague, and kept repeating the same questions, not seeming to recall her answers. She took him to the emergency department of their local hospital. He was seen by a neurologist and all the investigations came up as normal. He was diagnosed with transient global amnesia, given a sedative and admitted to hospital for the night. When he awoke the next morning, he had fully recovered. We don't know if he ever told his wife about his affair, or if he ever visited his lover again.

AN ORGASM IS A UNIQUE EXPERIENCE FOR YOUR BODY and, of course, for your brain. It is considered the peak or 'climax' of sexual experience and is associated with specific physical and psychological changes. These changes were first documented by William Masters and Virginia Johnson, groundbreaking sex researchers at Washington University in the United States. In the late 1950s, they conducted a series of experiments to document the physiological changes that occur during sexual activity. The TV series *Masters of Sex* portrays their revolutionary research, which initially faced many barriers. In a scene from the pilot episode, Masters presents his proposal to study what happens to the body during sex to his boss at the hospital where he works as an obstetrician and gynaecologist. His boss tells him he is jeopardising his career to study such a risky subject. 'It's smut, Bill,' he tells Masters, adding that the hospital board will label his proposal as 'pornography or prostitution or something equally depraved'. The study 'will never be seen as serious science,' he adds, 'and you will be labelled a pervert'. Masters laments the lack of support in an eloquent monologue:

> Every museum in the world is filled with art created
> from this basic impulse – the greatest literature, the
> most beautiful music. The study of sex is the study
> of the beginning of all life, and science holds the key.
> Yet we sit huddled in the dark like prudish cavemen,
> filled with shame and guilt, when the truth is: nobody
> understands sex.

Some of this context represents creative licence on the part of the series' producers, but in reality Masters and Johnson persisted with their novel research studies, which involved directly observing people masturbating, or having sex with randomly assigned partners, while wired up to various monitors to measure things like heart rate and skin conductance. Some participants even had a camera inserted intravaginally to photograph changes that occurred within the vagina during orgasm. Masters and Johnson participated in the study themselves; they became lovers and eventually married.

Masters and Johnson outlined a 'sexual response cycle' that comprises four distinct phases: (1) *excitement*, an initial state of arousal that involves increased heart rate, blood pressure and muscle tone (tightening of muscles); (2) *plateau*, or full arousal, which immediately precedes orgasm and is associated with a further increase in muscle tone, heart rate, blood pressure and breathing rate; (3) *orgasm*, or sexual climax, during which the muscles spasm, and heart rate, respiratory rate and blood pressure all peak; and (4) *resolution*, which involves the normalisation of all the physiological changes associated with orgasm. More recently, researchers have criticised Masters and Johnson's model for its failure to consider psychological, relationship and cultural factors, and for its assumption that people progress from one phase to the next in succession, which may not be the case. Other researchers have also revised the sexual response cycle. For example, an additional phase of 'desire' was introduced in the late 1970s by sex therapist Helen Singer Kaplan, who proposed a three-phase model of desire, excitement and orgasm. Our understanding of the human sexual response, however, is still largely thanks to Masters and Johnson's trailblazing work. Despite the criticisms and variations proposed in alternative models, there is no disputing the fact that an orgasm is widely regarded as the peak of human sexual experience.

Masters and Johnson were more interested in the physical responses that occurred during sex than what went on in the brain, but they did use scalp EEG – where electrodes are placed on the scalps of participants – in their experiments, which would have provided them with crude measurements of the brain's electrical activity. Nevertheless, this method is plagued with problems, such as the disruption of measurements by body movements, and later studies have found no specific changes in scalp EEG measurements during orgasm. The advances in neuroimaging technologies since Masters and Johnson's time now allow us to see what happens in the brain during orgasm, and several studies have examined this very issue. The brave participants in these studies have managed to have an orgasm, either self-stimulated or with a partner, while lying in the claustrophobic tunnel of an MRI machine or PET scanner, being observed by scientists and technicians as the patterns of blood flow and glucose metabolism in their brains are recorded – all for the sake of science. Most studies of orgasm in men have used PET scanning, which has the limitation of acquiring data over one to two minutes; this is too long to capture the relatively short event of ejaculation, and means that the brain imaging findings relate to events that occur immediately before and after orgasm (such as high sexual arousal and sexual satisfaction post-orgasm). Nevertheless, some studies of both men and women have documented a deactivation of prefrontal brain regions during orgasm, interpreted as the disinhibition or 'lack of control' required for orgasm to actually occur, while more recent studies have found activation of these brain regions in the lead-up to and during orgasm (see Chapter 2). These differences in findings are thought to be due to methodological variations in neuroimaging methods, and whether the orgasm was self-induced or via partner stimulation. There is still more work to be done to fully understand the neural correlates of human orgasm.

I spoke with the only researcher in the world who is currently doing empirical studies of the physiology of human orgasm. Dr Nicole Prause is a psychologist, neuroscientist and sex researcher who started studying the psychophysiology of sexual behaviour when she was an undergraduate student at Indiana University, where the famous Kinsey Institute is based. She uses various standard psychophysiological measures, such as skin conductance and heart rate monitors, in addition to her purposefully designed anal probe to measure contractions during female orgasms. She is currently working on a wireless intravaginal device. 'I never have any problem recruiting participants for my studies. People love them! The only exception is that they don't like the anal probe. If they don't want to come in, that's usually why.'

Despite sticking with the topic of sex for her entire research career, she admitted that the focus comes with some drawbacks. 'In hindsight, I would not have focused so much on sexuality, because the controversy around it really has been a career struggle. But the area is obviously fascinating, and it comes with a balance of fun and controversy. It lets you be cutting edge when others are too chicken to do it.' It is clear from Nicole's experiences that the stigma and controversy around human sexuality research remains strong. Masters and Johnson struggled with institutional constraints in their sex research more than 60 years ago, and yet Nicole is still facing the same battles. 'There's a reason other people aren't doing it. There's challenges with getting grant funding, and even when you get it, sometimes Congress rescind the funding from scientists studying this stuff.'

In the early 2010s, when Nicole was a research scientist in the Department of Psychiatry at UCLA, she applied to the ethics board to do a research study on female orgasm. The board would not approve the study. When she was awarded a grant from a non-profit organisation to study 'orgasmic meditation' in couples and the university would not allow her to accept it, she realised she had to leave the university if she wanted to continue in the field: 'Clearly this research cannot be done at a US university campus – it's just too controversial!' So, out of necessity, Nicole started her own research lab, *Liberos*, where she currently conducts her own studies, free of the hassles of university administration. She currently collaborates with other academics from various universities. 'This works really well,' she said. 'It means their universities can outsource the controversy to me.'

Nicole told me that there is very limited knowledge about basic orgasm physiology, particularly in women. Half her female sample in a recent study self-reported having an orgasm but didn't actually show the physiological response of contractions that accompany orgasm. She doesn't know what to make of this and wishes there was more understanding 'just about the basics of orgasm'.

To avoid some of the limitations of early scalp EEG technology, Nicole uses more modern EEG technologies in her studies to look at brain responses during orgasm. She said EEG is useful because of its speed, as you only get one trial with orgasm research. 'I'm only able to do it because of the new bluetooth EEG headsets, as they allow movement.' The trade-off, she said, is that 'you can't do localisation', which would identify the specific brain regions the electrical changes occur in: 'You can only really look at the whole brain response.' Obviously people move quite a bit when they masturbate and orgasm, so the old 'bite bar' to help subjects keep their heads still is not an option. Her research group attaches an accelerator to the back of their heads to monitor movement, so this can be mathematically removed from the data for analyses.

She told me about some interesting EEG results from her research on female orgasm. Women are given an hour to reach orgasm in the lab. Nicole has found that just *before* her subjects are told that 'they can go for it', they show a strong suppression of alpha wave activity, a type of electrical brain activity that you see when people are awake and alert. They also show an increasing galvanic skin response, which is a measure of sweat on the skin; an increase is associated with emotional arousal. But then something intriguing happens: 'As soon as they are told they can go for it, this flips – we see a precipitous drop in galvanic skin response, and an increase in alpha activity.' I ask Nicole why she thinks that's the case:

> What we think is happening is that you need to reduce neural inhibition to allow enough synchronous neural firing in order to trigger a climax, or to trigger whatever it is that triggers an orgasm. We still don't actually know what that is, but we think it might be similar to some seizure disorders, and that there has to be some amount of synchronous neural firing to allow the rhythmic muscle contractions to occur in the pelvis. There's probably a need to release neural control to allow orgasm to occur.

I was surprised to hear yet another link between sex and seizures! Nicole's interpretation of her EEG results echoes some previous neuroimaging findings of deactivation of prefrontal brain regions during orgasm. Her study is currently under peer review and yet to be published, but it will no doubt make an important contribution to the scarcely researched topic of the neuroscience of female orgasm.

Nicole feels it's important to emphasise the broader implications of sex research. 'Sex research is never going to be important for those people who think sex is only for procreation,' she pointed out. 'We need to show it has a health application.' She is exploring how orgasm can be used to facilitate sleep, and how work productivity may be related to sex, by doing cognitive tests with her participants before and after their orgasm or sex research sessions. Is it reasonable to masturbate at work if it means you'll be more productive after it? Grant funders and university ethics committees might not care about the intensity of orgasms, but if sex research can be related to these broader implications for health and productivity, the path to getting this research done may be a lot smoother in the future.

AS WE HAVE SEEN, THE SUDDEN BLOOD FLOW CHANGES that occur in the brain during orgasm can occasionally trigger an episode of transient global amnesia. There are also other neurological events that can be triggered by sex. Sex can induce seizures (see Chapter 2) and even stroke. If you are unlucky enough to have a brain aneurysm, a malformation that looks like a small balloon protruding out the side of a blood vessel in your brain, intense physical activity such as sex (especially with an orgasm) can cause the aneurysm to rupture or burst. A ruptured brain aneurysm can result in blood leaking into your brain, which is called an *intracranial* haemorrhage and is a type of stroke, or it can cause a *subarachnoid* haemorrhage, which means that the blood has leaked under the arachnoid mater, one of three membranes that covers the brain.

A brain aneurysm can grow anywhere within the blood vessels of your brain, and the effects of a ruptured aneurysm depend on where it is located. Why they appear is unknown, but certain risk factors such as smoking, high blood pressure, and heavy alcohol or cocaine use can increase your chances of growing a brain aneurysm.

Sometimes it is just the luck of the draw of inherited genetic conditions. They are more common in women than men. It is estimated that one in 50 people have an unruptured brain aneurysm, and most people would not even know they have one – they are most often discovered incidentally during a brain scan in search of other conditions. Treatment of an unruptured brain aneurysm to prevent it from rupturing and bleeding involves 'clipping' (where the blood vessel under the aneurysm is clipped), endovascular treatments (coiling, stents or flow diversion), or just leaving it alone and monitoring it over time. A ruptured brain aneurysm causes death in 40 per cent of cases; of those who survive, two-thirds end up with a neurological deficit. This is what happened to Samantha.

Sociable 25-year-old Samantha had no history of psychiatric illness. Considered a 'party animal' in her circle of friends, she had drunk alcohol heavily since her teenage years. Her friends described her as 'impulsive', 'promiscuous' and always popular with the boys, but she had eventually settled into a relationship with Cameron, her on-again-off-again boyfriend of seven years, and they planned to marry. One evening, while having sex, Samantha suddenly developed an extremely severe headache, nausea and vomiting. She was found to have a haematoma – a bleed – in her right temporal lobe from an arteriovenous malformation (AVM), a tangle of abnormally formed blood vessels that are prone to bleed. A neurosurgeon operated on her to evacuate the blood and remove the AVM. After the surgery she had left-sided hemiparesis (weakness on the entire left side of her body) and a left homonymous hemianopia (impaired vision in her left visual field).

Three months after her surgery, Samantha developed seizures that arose from her right temporal lobe; these can occur after a bleed in the brain – an intracranial haemorrhage – due to the blood causing scarring of brain tissue. Her behaviour also changed, and she was described as 'bizarre' and 'not her normal self'. She became 'a nuisance to her neighbours' and would collapse in their doorways; she was aggressive towards her mother, who struggled to care for her. She frequently turned up at the emergency department of her local hospital claiming to have had another brain haemorrhage. Cameron couldn't cope with this new Samantha and left her.

Samantha had had the disastrous experience of a sex-induced ruptured AVM, which was followed by seizures arising from her right temporal lobe – and, as we know, this is a crucial part of the sexual neural network (see Chapter 1). Her neurological condition later deteriorated further when she developed erotomania, a love delusion, as we will see in the next chapter.

So, if you have a brain aneurysm, how likely is it that having sex will cause it to burst? One study estimated that intense physical activity such as sex can make an aneurysm up to 15 times more likely to rupture. Case studies have reported that sex was the immediate preceding activity before a ruptured aneurysm in up to 14 per cent of patients, and this is much more common in men (33 per cent) compared to women (7 per cent). This gender difference is a surprise: brain aneurysms are actually more common in women, and women have longer orgasms than men. It has been proposed that the higher risk of sex-induced subarachnoid haemorrhage due to ruptured aneurysm in men is due to the greater increase in arterial blood pressure that they experience during sex.

How can sex cause a brain aneurysm to burst? The two key factors that make brain aneurysms rupture are increases in blood pressure and a sudden reduction in intracranial pressure. This is exactly what occurs during sexual activity, particularly in the late plateau and orgasm phases of the human sexual response cycle outlined by Masters and Johnson. The profound changes in blood pressure, heart and respiratory rate, and muscle tension that accompany these sexual response phases result in increasing pressure on the wall of the aneurysm, making it more likely to burst. This is a highly dangerous event; many die before reaching hospital and there is significant disability in the majority of survivors. When it is triggered by sex, it turns out that sudden death is much more likely if you are having sex with someone who is not your usual partner. In other words, the excitement of extra-marital or 'unfaithful' sex in an unfamiliar environment is associated with more intense cardiovascular changes (that is, to heart rate and breathing rate) than those that occur during sex with your long-term partner in your bed at home. Heavy food and alcohol intake are also associated with increased blood pressure and the risk of ruptured brain aneurysm. So, if you have a brain aneurysm, think twice before booking that fancy restaurant and hotel room for a fling. Sticking with your current partner might just save your life.

To answer the question posed in the title of this chapter, yes, sex *can* actually trigger a dramatic change in your brain, and on rare occasions can even damage your brain. Where you have sex, who you have it with and whether you reach orgasm all have an impact on exactly what happens in your brain. These are fascinating observations that raise more questions than answers.

6

WHAT'S LOVE GOT TO DO WITH IT?

Barry kicked off his dirty workboots by the front door. He grunted a greeting at his wife, Sue, as he made a bee-line for the fridge, his mind focused only on the cold beer that awaited him. Sue sighed as she heard the creak of his recliner and knew that she was facing another night of the same old boring routine – of trying to engage her husband in conversation while he grunted replies at her, his eyes fixed on the TV screen.

It wasn't what she had imagined their life would be like when they had met one hot New Year's Eve in their twenties, when they had both laughed about their matching sunburned noses and jokingly called each other 'Rudolph'. She wasn't sure how much longer she could stand it. Barry was a grumpy and ungrateful husband. She tried to remember the last time he had made eye contact with her, let alone when they had last laughed together. Maybe it was when he gave her a kiss on the cheek just before he left on an interstate work trip a few months ago?

In an instant, that would all change.

Later that night, Sue woke to find Barry crying out in pain and clutching his head. He had suffered a stroke – to be precise, a subarachnoid haemorrhage due to a ruptured left posterior communicating artery aneurysm. Brain scans showed damage to the left hemisphere of his brain, in particular his left temporal lobe and basal ganglia, which had been scarred by blood and starved of oxygen.

I first met Barry and Sue three months afterwards. Surprisingly, considering he had experienced a major stroke, Barry said he did not have any complaints except for some occasional word-finding difficulties. He smiled and told me he felt great. 'Better than ever,' he said. 'It's weird!' When I asked if I could speak with Sue alone, he kissed her on the hand and said, 'See you soon, my love.' She laughed and looked shyly at me.

After the door closed and we heard his footsteps fade away down the hallway, Sue revealed the colossal shift in their relationship. 'It's marvellous how this has changed him,' she told me. 'He's done a one-eighty! He's happy all the time, enjoying life, laughing more. He loves everything now. He's different – it's a new version of the old Barry. He's always saying how much he loves me, and "You're beautiful." It's wonderfully overwhelming.' Barry had unexpectedly transformed into the romantic husband she'd always hoped for, a man brimming with joy and constantly professing his love for his wife. His burst brain aneurysm had saved their marriage.

It is rare for anyone affected directly or indirectly by a

brain injury to consider it a positive thing, but if it resulted in enhanced love, and a more loveable version of the person you knew before, who wouldn't consider it a blessing? Barry's case raises obvious questions. Are there certain brain regions that are crucial for love? Were these regions somehow 'released' by Barry's subarachnoid haemorrhage? Are these parts of the brain the same as the ones that are involved in the sexual neural network? In other words, do the same parts of the brain influence sex *and* love?

THE FEELING OF ROMANTIC LOVE – FALLING IN OR OUT of it, being in it or losing it – is a distinctive and unforgettable human experience. Love is one of the most common themes in songs, films and literature. Songs tend to refer to the heart as the organ of love, but really we should be singing about the brain. I guess singing about having your brain broken, or brainache, or a burning love that springs from the bottom of your brain doesn't quite have the same effect.

Love can consume and devastate us; it can bring about life or make us want to destroy it in a jealous rage. It is a universal emotion, thought to have existed across all cultures since humans first evolved. Biological anthropologist Helen Fisher, a leading researcher in this area, proposes that romantic love evolved to motivate our ancestors to focus on and conserve energy for a specific mating partner, leading to attachment and the benefits of mutual parenting. She considers love to be a natural 'addiction' and highlights how people who are in love or rejected in love show the same symptoms as those with substance (such as amphetamines) and behaviour-related (such as gambling) addictions. Craving, mood changes and, in the case of a break-up, withdrawal sypmtoms are common to all these addictions.

Fisher's theory that equates love with addiction is supported by neuroimaging studies that have examined the brains of people in love while they look at pictures of their romantic partner. Using functional MRI brain scanning techniques (fMRI), which measure blood flow, researchers have identified brain networks related to romantic love that overlap with parts of the brain involved in our neural 'reward system'. Our brains respond to rewards by producing the neurotransmitter dopamine (see Chapter 2), and the dopamine neural pathways are part of the reward system. The mesolimbic pathway connects the ventral tegmental area (one of the main dopamine-producing structures) with the ventral striatum of the basal ganglia; this, in turn, has extensive links with frontal brain regions. Brain scans of people in love have revealed activation or increased blood flow to parts of the frontal lobe (such as the anterior cingulate cortex), basal ganglia (nucleus accumbens and caudate nucleus) and hypothalamus – brain regions that not only process reward but are also associated with substance and behavioural addictions, including the controversial porn and sex 'addictions' (see Chapter 9). This overlap provides support for Fisher's 'love as an addiction' theory.

The regions activated in these loved-up brains also overlap with many crucial parts of the sexual neural network, such as the amygdala and prefrontal cortex, but both romantic love and sexual neural networks are widespread and involve structures throughout the brain. Brain imaging can only reflect a snapshot of time, and cannot possibly capture the complex experience of sex or love. It can, however, provide some intriguing insights into the brain regions that are involved in both.

Amazingly, different patterns of activation and functional connectivity between brain regions have been found when comparing people in and out of love, and those who have never been in love. Even when the brain is 'at rest' or 'in neutral' – that is, not engaged in a particular task (such as staring at a photo of a lover) – there are differences in the brains of people who are in or out of love. A part of the frontal lobe (specifically, the left dorsal anterior cingulate cortex) shows differences in 'regional homogeneity', or synchronised neural activity, depending on whether you are in or out of love, and how long it has been since you experienced love. People who are 'in love' have increased left dorsal anterior cingulate regional homogeneity or neural activity that is more 'in synch'. Surprisingly, this is positively associated with the length of time they have been in love, so it is greater in those who have been in love for a longer period of time than for those experiencing the first flush of new love. On the other hand, for those out of love, the longer it has been since a break-up, the *less* regional homogeneity will be found in left dorsal anterior cingulate cortex.

Furthermore, when people are in love, their brains display increased functional connectivity – that is, an increase in spontaneous blood-oxygen-dependent signals between spatially remote regions and networks of the brain. When we are in love, this functional connectivity increases between the regions that are involved in reward, motivation and emotion regulation (such as the amygdala), and in social cognition, or the types of thinking skills required to negotiate social relationships (such as the medial prefrontal cortex and temporal lobe). So in a 'chicken and egg' kind of way, it seems that love enhances the patterns of activation within brain regions that control the very functions that you need to feel love! These research findings suggest that a brain scan might be able to prove if people really are madly in love, or if they have never really been in love at all.

Patterns of brain activity that occur early during a period of romantic love can even be used to predict whether a relationship will last. In one study, people experiencing the early phase of romantic love had fMRI scans; then, 40 months later, they were rescanned. The brain activation patterns of those who were still in their relationship were compared with the patterns in those who had broken up. People who were still with their partners actually showed *less* activation in their first scans in frontal regions of the brain (specifically, the medial orbitofrontal cortex and right subcallosal cingulate) and the basal ganglia (right nucleus accumbens) when compared with those whose relationships had ended. Incredibly, they found the less activation in those brain regions early in the relationship, the greater their relationship satisfaction later in the relationship. So there is preliminary evidence from this study that your brain activation early in your relationship can predict your relationship stability and satisfaction 40 months later. If your brain activation is through the roof in the early stages of getting together, it may be an omen for the future of your relationship!

The neuroimaging studies of the brain in love discussed so far have been 'functional' studies of blood flow and glucose metabolism. There has also been a study of the neural implications of love that has looked at the structure of the brain. Compared with single people, those in the first month of a romantic relationship have been found to have reduced grey matter density, which reflects the loss of certain parts of neurons, in the right 'striatum'. This is part of the basal ganglia and includes structures, such as the caudate nucleus, which are part of the reward neural network (see Chapter 9). People in a relationship scored higher on a subjective questionnaire of happiness, and the authors suggested that this positive experience of being in the early stages of a romantic relationship could lead to structural brain changes. Nevertheless, this study cannot tell us about the causal connections between these two factors. It may be that people with reduced right striatum grey matter density are just more likely to have a romantic relationship, rather than the romantic relationship causing this brain change. To investigate causal changes of romantic love, we need a longitudinal study involving brain scanning before and after a romantic relationship. Maybe researchers should start advertising on Tinder to find willing participants ...

But what about the brains of people who have fallen out of or been rejected in love? The idea of a 'broken heart' makes direct reference to the fact that a romantic break-up can cause physical pain. Losing love is a feeling that most people will experience at some point in their lives and never forget. Like grief and depression, it can cause a range of physical symptoms, from reduced appetite to sleep difficulties. Brain imaging studies of people who are 'broken-hearted' have found that romantic rejection and physical pain activate the same brain regions, specifically the anterior cingulate cortex and the anterior insula. One study found that these brain regions were activated when people viewed pictures of a person who had recently broken up with them, and when they were given painful heat stimulation. Therefore, the experience of romantic rejection and physical pain are underpinned by the same neural regions. 'Heartbreak' might be an anatomically flawed idea, but it does reflect the physical pain that people can experience when love comes to an unhappy end.

IT IS NO SURPRISE THAT THE NEURAL NETWORK UNDER-lying romantic love is as complex as love itself. Love involves the full array of human emotional and physical experiences, so it makes sense that it is mediated by widespread brain regions that overlap with the sexual, reward and pain neural networks. We can also learn about brain regions that are involved in love from two fascinating types of delusions that can occur in the context of brain injury or disease:

erotomania and Capgras delusion. Erotomania, also known as De Clérambault's syndrome (after the French psychiatrist who first described it), is the sudden belief that someone of higher social status has fallen passionately in love with you, and that you are deeply in love with them. It typically manifests in middle-aged females with psychiatric illnesses such as schizophrenia or bipolar disorder, but there are rare reports of it occurring after an acquired brain injury or in people with dementia. For Margaret, it was the first symptom of a sinister problem.

Margaret was 65 years old. She had two children, and had lived alone since her divorce. Her medical history included a transient ischaemic attack – a brief and temporary interruption of blood flow to the brain. These are often a warning sign that you are at risk of having a stroke. She had been a heavy smoker, working through up to 40 cigarettes a day, but had quit five years before she developed the psychiatric and neurological problems that brought her to the attention of the authors of her case study.

Over a period of six months, Margaret became increasingly obsessed with a deceased American singer of the 1940s and '50s, Dick Haymes, who was famous for his love ballads and his marriage to actress Rita Hayworth. Margaret bought every CD and DVD she could find that featured Haymes, and his music was a constant soundtrack to her life, broadcasting across her entire house from devices in all rooms. Despite being told he was dead, she insisted the singer was alive, and believed he was living next door to her and was in love with her. She said she 'loved him to bits' and on one occasion she waited for hours outside her home, anticipating his arrival. When her daughter challenged her about her apparent delusion, Margaret slapped her in the face, something she had never done before.

Her daughter discovered her mother had not been eating or sleeping properly, and a neighbour told her he had seen Margaret wandering the streets late at night. She insisted on taking Margaret to the doctor. The doctor sent her for an MRI brain scan which revealed chronic small vessel ischaemia – brain changes related to a lack of oxygen to the small blood vessels in the brain, often found in heavy smokers – and atrophy, or shrinking, of the anterior (front part) of the temporal lobes. A brain scan called a SPECT (single photon emission computed tomography), which measures glucose metabolism, showed hypoperfusion, or reduced metabolism, in the left frontal and temporal regions. This means that those brain regions were not functioning properly and, as we know, those regions control a vast array of functions that essentially make us who we are – and they are also considered part of the sexual and love neural networks. Her neurological condition had left Margaret unable to interpret one of our most complex social behaviours – love.

She was started on antipsychotic medication and her erotomania resolved, but other symptoms started to appear. Her speech became slurred and soft, her tongue twitched intermittently and she had difficulty swallowing. The muscles in her hands wasted away, and she had to abandon knitting, her favourite hobby, as she couldn't hold onto the needles anymore. She lost weight and her legs became weak. Eighteen months after she had been seen for her erotomania, she was sent for an urgent neurological review. During her appointment, she was impulsive and demonstrated 'utilisation behaviour' – or a tendency to use any object placed in her vicinity, like a baby grabbing at toys on their highchair table. This is indicative of severe frontal lobe dysfunction. She grabbed the neurologist's pen from his desk and started writing on his notepad, and leaned forward and reached for the glasses that protruded from his shirt pocket. During a neuropsychological assessment she had to be physically restrained to stop her from grabbing at objects on the desk. She had difficulty concentrating and could only comprehend very simple sentences. After a series of investigations, she received a diagnosis of motor neurone disease with frontotemporal dementia. Motor neurone disease was initially thought to only affect nerve cells controlling the muscles, but it is now known that it can also cause changes to cognition and behaviour in up to half of the people who are diagnosed with the condition. In some cases like Margaret's – an estimated 5–15 per cent of those diagnosed – these changes will be severe enough to warrant a diagnosis of behavioural-variant frontotemporal dementia as well.

In Margaret's case, erotomania was the first sign of the changes that were occurring in her brain as a result of a tragic neurodegenerative condition. In the case of Samantha, introduced in Chapter 5, erotomania came three years after her right temporal lobe was damaged by an intracranial haemorrhage that occurred during sex. Samantha was by this time 29 years old, and she had begun to experience paranoid delusions of great concern. One was that people were spying on her and trying to kill her. Another was her passionate desire and love for her real estate agent, from whom she had rented a flat. She believed he was in love with her and wanted to marry her, but that he was prevented from doing so by the government. He maintained that all he had done was organise her lease. She stalked him, sending numerous letters and calling him hundreds of times, to the extent that he moved to a different city to flee her persistent attention – only to find that she had managed to track him down again. The case report stated that Samantha's paranoid delusions resolved with medication, but her erotomania remained unchanged. There is no other information about what happened to Samantha. Her case is a tragic and very rare example of how sex can change your brain, and your brain in turn can alter your sex and love life forever.

Another group of delusions, called 'misidentification delusions', involve the belief that a familiar person, object or place has been altered. These types of delusions typically occur in psychiatric conditions such as schizophrenia, and in neurological disorders including Alzheimer's dementia. A dramatic example of this is Capgras delusion, when a person believes that a significant other – typically a spouse – has been replaced by an identical impostor. In other words, their feeling of familiarity or love for the person vanishes. It could be characterised as a sudden falling out of love, the flip side of erotomania.

Barbara was diagnosed with Alzheimer's dementia five years before I met her. She would often get confused and agitated. On some occasions, she would even fail to recognise her husband, David. It would occur suddenly and out of the blue. For example, if she left David to go to another room, on her return she would ask him, 'Who are you? Where's David?' and order him out of the house. These episodes typically lasted between one and three hours. Her husband would go outside, sit in his car and wait. On one occasion, she said, 'Don't go and sit in the car – that's David's car.'

Whenever this happened, David had no idea how to make Barbara understand that he was her partner of nearly 60 years. He told me that after this first occurred, he decided to sleep in a different bed, as he did not want to frighten her. No doubt it would have been terrifying for both of them if she had awoken in the night, in the midst of a Capgras delusion, believing that an intruder was lying next to her. David now claims that it was the power of a love song that resolved this delusion and brought Barbara back to him. On the night they first met, they had danced to the last song of the evening, 'Unchained Melody'. He began singing this to her several times a day, and within a couple of weeks, he said, 'She came back. Everything changed. It worked quickly. She stopped sending me away, and became my shadow.' The episodes of her failing to recognise him stopped occurring. Barbara and David had heeded the lyrics: 'I'll be coming home, wait for me.'

Music is a powerful treatment tool for symptoms of dementia, and songs from people's younger days can take them back in time to their first loves, even if only for the duration of the song itself. In the midst of the multitude of challenges of caring for someone with dementia, such moments remind our loved ones who they were, and the relationships they have. With no cure for dementia, maximising these moments is crucial.

Just as our brains can change or be changed by sex, love also alters and is altered by our brains. Brain injury or disease can make us fall in or out of love in sudden and dramatic ways. We do not fully understand the complexities of love; likewise, there is still so much more to learn about love in the brain.

AFTER MONTHS OF REHABILITATION, BARRY EVENTUALLY returned to work. When he returns home each evening, the first thing he does is kiss Sue, ask how her day was, and tell her that dinner smells delicious. The days of grumpy pre-stroke Barry are long gone, and Sue is thankful for the change that gave her a loving husband. I never asked about their sex life, but I have no doubt it improved. His stroke caused extensive damage to the left hemisphere of his brain, including the left frontal and temporal lobes, which we know are part of both the love and sexual neural networks. The positive outcome of his brain injury could also be due to it affecting the left side of his brain, leading to an improvement in the expression and experience of all positive emotions, not just love. In a study of people with frontotemporal dementia, those with greater left frontal atrophy were found to show increased expressions of happiness in response to a film. The authors of this study proposed that selective damage to left frontal brain regions may weaken positive emotion regulation and correspondingly facilitate positive emotional responses – or expressions of happiness. This extends earlier research that suggested the left frontal lobe is involved in generating positive emotion, and indicates that damage to this area may actually enable positive emotions to be expressed.

Dustin is the only other patient I have seen with a severe brain injury who reported an exclusively positive outcome. He also had extensive left hemisphere damage, but his was the result of a very severe traumatic brain injury he had sustained in an assault. Before his injury, he had a long forensic and psychiatric history, including a diagnosis of antisocial personality disorder. Remarkably, when I spoke with him, he told me that his brain injury was 'a good thing': 'Before it, [I was] an arsehole. My kids hated me and I did bad things. Now they love me.' His mother also remarked, 'This injury was the best thing that happened to him. He's a much nicer person now.'

In the cases of both Dustin and Barry, their left-sided brain injuries enhanced their ability to express positive emotions, including love. This resulted in them being generally much nicer guys who were more pleasant to be around, enabling happier social relationships. Luckily, both expressed their newfound positivity and love in a socially acceptable ways. Sometimes, however, a brain injury or disease can release a sexual or love interest in something other than a consenting human partner, which can lead to devastating outcomes for the patients and their family members. These paraphilias are discussed in the next chapter.

7

FROM SAFETY PINS TO SLEEPING BEAUTIES

For as long as he could remember, Harry had loved safety pins. He found the shape, colour and shine of them so incredibly beautiful that just holding one gave him an intense feeling of pleasure. He called this special feeling 'thought satisfaction'. As a child he collected safety pins, and his pockets were always bulging with them. His favourite game was to join them in a long chain and pull them along the floor. He knew it was an unusual and potentially embarrassing habit, so he would hide in the toilet for his special safety pin time, pulling the pins out of his pockets one at a time and staring at them shining in his palm. Sometime during his childhood, the 'thought satisfaction' was followed by a 'blank period', which was eventually diagnosed as a seizure when Harry was an adult. Somehow he was able to keep this secret from his family, and the first witness to his safety pin habit was his wife.

Harry was 23 years old when his wife found him staring at a safety pin in his palm; he was glassy-eyed, vague and unresponsive. The seizures triggered by the safety pins always followed the same pattern: he would stare at a safety pin and start humming, make sucking movements with his lips, pluck at his clothing, and fall into an unresponsive state for a couple of minutes. Every seizure was triggered by a safety pin, and therefore they were essentially voluntary. It had to be a bright, shiny, undamaged pin, and several were more effective than just one. He had the strongest desire to look at pins during anxiety-provoking and sexual situations. If he fantasised about safety pins during sex, he would have a seizure. Over time, Harry lost all interest in and desire for sex with his wife, became impotent, and only lusted after safety pins. He described the 'thought satisfaction' triggered by safety pins as the 'greatest experience of my life, better than sexual intercourse'.

Harry tried all the anti-seizure medications known at the time – this was the 1950s – with no benefit. He had a routine EEG, which showed a left temporal seizure focus, and was referred for surgery. On 17 March 1953, he underwent a left temporal lobectomy. His surgery was considered a great success: it cured him of both his seizures and his safety pin fetish. The case notes from his review 16 months after his surgery state that he had 'no desire to look at safety pins and had become as potent as in early marriage', and that his relationship with his wife improved. It doesn't provide any more details about their relationship, pre- or post-surgery, but no doubt his wife would have been relieved. If Harry had been alive today, I wonder if our more advanced anti-seizure medications would have been more effective; he may not require a left temporal lobectomy to experience both a seizure and fetish cure now. Although this case study is over 60 years old, it remains extraordinary – for the unique nature of the fetish, its clear association with epilepsy, and its unequivocal evidence that the temporal lobe is a crucial part of the sexual neural network.

According to the current and fifth version of the *Diagnostic and Statistical Manual of Mental Disorders* (DSM-5), a paraphilia is an intense or persistent sexual interest in something other than 'genital stimulation or preparatory fondling with phenotypically normal, physically mature, consenting human partners'. Common paraphilias are *fetishism*, or a preference for particular body parts or inanimate objects (such as Harry's safety pins); *voyeurism*, which involves spying on others in private activities; and *exhibitionism*, or exposing genitals. Rarer and less known paraphilias include *symphorophilia*, or sexual pleasure from watching a disaster such as a car accident (as featured in JG Ballard's novel *Crash* and its Oscar-winning 1996 film adaptation); *formicophilia*, or having insects crawl on your body; and *infantilism*, or dressing up or wanting to be treated like a baby, which can also involve *diaperism*, or deriving sexual pleasure from wearing a nappy.

Another rare paraphilia is *somnophilia*, also known as 'sleeping beauty syndrome'. It has been represented on screen in the films *Sleeping Beauty* (2011) and *The Little Death* (2014), an Australian production that also portrays several other paraphilias. In a case study published by Swiss psychiatrist Francesco Bianchi-Demicheli and his colleagues in 2010, a link was made between this paraphilia and a childhood brain injury. I read it and couldn't help but say 'Wow' out loud to my empty office.

Jim, a 34-year-old man, arrived at an emergency psychiatric unit after physically assaulting his wife. His relationship with his wife had been strained for several years, because over time he had developed 'a particular and progressive sexual deviant behaviour': he felt sexually aroused by sleeping women, and liked to attend to their hands and nails while they slept and give them a manicure. During the initial years of his marriage he was able to control these fantasies but, the case study reports, 'over the years he lost control of his sexual urges' and acted on them. Initially Jim's wife agreed to use sleeping pills to fulfil his desires, but she later refused to 'bend to [the] man's freakish will'. That's when he began to secretly administer benzodiazepines – minor tranquillisers, typically prescribed to relieve anxiety and insomnia – to her. When Jim's wife discovered what he had been doing, their conflict began. It all came to a dramatic head when Jim 'disguised himself with a latex mask and attacked his wife as she returned from work', using a capsicum spray to subdue her. His wife managed to pull off his mask, then ran out of the house and called the police.

The results of the psychiatric examination and routine laboratory tests were all normal. The only medical history of note was a traumatic brain injury Jim had suffered at aged ten years, after which he was reportedly in a coma for four days. It doesn't say how he sustained this injury, but an MRI brain scan showed that it caused damage to his right frontal and parietal regions, specifically 'moderate atrophy in the frontoparietal region with a diffuse and severe white matter injury compatible with his previous head trauma'. A neuropsychological assessment was said to show 'moderate dysexecutive syndrome', which means he did not do so well on the tasks that examine frontal lobe functions, such as response control.

We know the complex array of social and emotional functions the frontal lobes are involved in (see Chapter 2), but what do the parietal lobes do? They process sensory data that enable us to understand spatial information, including where our body is in space – in other words, 'body awareness'. People with parietal lobe damage can have abnormalities in how they perceive their bodies, which is exactly what Jim showed: he had a fascinating and very specific body disorder characterised by an incomplete mental image of his hands, mostly on the right, called 'personal representational hemi-neglect'. This was determined by the way he drew a picture of himself – he neglected his hands. The authors concluded that Jim's paraphilia 'might be related to his disturbed body image and more specifically to the incomplete hands representation ... presumably the occurrence of head trauma might have played a critical role in the development of his sexual self- and body image'.

If a paraphilia causes a person distress in everyday life, impairs them in some way, or puts them or others at risk of harm, it is considered a 'paraphilic disorder'. Therefore, the term *paedophilia*, or the sexual preference for prepubescent children, is distinguished from a *paedophilic disorder* –

that is, paedophilia becomes a disorder if these sexual urges have caused distress or interpersonal difficulties, or if the individual has acted on those urges. As a team of neuroscientists stated in a recent review, 'the mere sexual preference for children has been depathologised'. This means that no therapy or prosecution is required for paedophilia that exists *only* as a preference or desire and results in no actions, impairment, suffering or harm. The authors of the review do note, however, that a paedophilic sexual preference may be a risk factor for later criminal offending, which warrants preventative measures. So if the desire leads to any sort of action, such as downloading child pornography or any type of sexual offence against a child, then it's a disorder, and that action is a criminal offence. Preventing these desires from becoming actions and offences is one of the main goals of research into paedophilia.

Regardless of the diagnostic term that is used, a sexual preference for children that exists as a thought or desire only is considered atypical and abhorrent, and there is no doubt that acting on it in any way is a criminal offence. A frightening fact is that there are rare cases in which this paraphilia has developed suddenly and unexpectedly in people with brain injury or disease. These cases provide the most striking evidence for the sexual neural network, particularly those cases in which (1) the treatment of the neurological condition (which in some cases involves the neurosurgical removal of the 'offending' brain region) results in the cessation of the paraphilia; and (2) the neurological condition returns, and the paraphilia returns with it. The extraordinary case of Gary, who developed paedophilic disorder associated with a right frontal lobe tumour, offers some insights.

Before reading Gary's case, please be aware that the text to come may be distressing to some people. Paedophilia is a very confronting issue. There are two main reasons why it is discussed here. First, the purpose of this book is to highlight cases in which people with brain injury or disease have experienced changes in their sex lives, and there are cases in which there is a clear association between a neurological condition and paedophilia or a paedophilic disorder. Such cases offer critical evidence for the role of specific brain regions in the sexual neural network. Therefore, from a scientific perspective, these cases need to be discussed. Second, we can only understand and ultimately prevent paedophilia through research and discussion. Failing to discuss it does not help victims. The goal of research into this condition is to protect future victims by learning what triggers offending, and how that can be prevented well before it happens.

GARY WAS A HEALTHY AND HAPPILY MARRIED 40-YEAR-old schoolteacher when he developed an increasing interest in pornography. He had enjoyed browsing it since his teenage years, but this was much more intense. He began obsessively collecting porn magazines and viewing pornographic videos online. He hid the magazines in the bottom drawer of his filing cabinet, but when this became full, he had to find a new hiding spot in the garage to conceal them from his wife and stepdaughter. The other difference in his new-found obsession was that he desired child pornography, something he had never been interested in before. He was actively searching internet sites for images of children and adolescents. He knew it was wrong, but he felt that his sexual impulses were just too strong to resist.

Gary developed a habit of visiting prostitutes and attending 'massage parlours' regularly. He'd pretend he was going to the gym, and even sneak in visits after his teaching duties were done before his wife got home from work. But it was when he started making sexual advances towards his prepubescent stepdaughter that things took a catastrophic turn. After several weeks, the child finally told her mother. Gary's wife was horrified. When she also discovered his porn stash and frequent access to child pornography on their family computer, she realised she had no idea who her husband was now. She was too shocked to grieve their marriage and focused on protecting her daughter. Gary was legally removed from their home and found guilty of child molestation. There's no further information in this case study about whether he pleaded guilty or not, nor the exact nature of the charges. Cases like Gary's raise complex questions about criminal responsibility and punishment that the legal system must grapple with; I discuss this further in Chapter 10.

During his sentencing hearing the judge told Gary that he could either go to jail or complete a 12-step inpatient rehabilitation program for sex addiction. Not surprisingly, he chose the latter option. Nevertheless, his sexual urges still were too strong for him to resist. When he began to ask staff and other clients at the rehabilitation facility for sexual favours, he was expelled from the program. The night before he was due to attend his prison sentence hearing, Gary developed a severe headache and attended a hospital emergency department. Given the timing, his headache was initially dismissed as stress-related, but when he discussed thoughts of suicide and a fear that he would rape his landlady, he was admitted to the psychiatry service. The next day, he complained of poor balance, and was referred to a neurologist. During the neurological examination he urinated on himself and showed no concern about it. He asked female doctors in the team for sex. He swaggered from side to side when instructed to walk, and had trouble writing and drawing a clock face, which is a standard cognitive test. It was more than enough evidence that something sinister was going on in his brain, and he was sent for an MRI brain scan.

The scan revealed a large tumour that was squashing his right orbitofrontal lobe. He was sent to the neurosurgical team and had the tumour removed. The tumour was found to be a particularly aggressive type called a hemangiopericytoma. After his operation, Gary's walking and bladder control improved, and his paedophilic behaviour and excessive interest in pornography resolved. He successfully completed a rehabilitation program, and seven months later he was not considered a threat to his stepdaughter. Amazingly, the case report notes that he 'returned home'.

There is no information about the state of his relationship with his wife.

Three months later, Gary developed a persistent headache and secretly started collecting pornography again. An MRI scan showed that the tumour had regrown. It appears that this regrowth of his tumour had switched his sexual impulses back on. He underwent a second neurosurgical resection of the tumour two years after his first surgery, and after that there is no further follow-up. I emailed one of the authors of this case study to ask for more information about Gary, his wife, and what happened after his second surgery, but received no reply. The case study was published in 2003, and the media went wild when it was presented at a conference; I read online that the authors were inundated by requests from defence lawyers who wanted help defending clients charged with paedophilia-related offences. So perhaps the authors were tired of the attention, and of being associated with the 'blame my brain for paedophilia' phenomenon (see Chapter 10).

Paedophilia is never the sole symptom of a neurological condition. It always occurs in the midst of other cognitive or physical changes, and typically in the context of hypersexuality, as in Gary's case. Paedophilia has long been considered a lifelong trait, a stable and unchangeable condition, and the cases in which it arises in the context of a brain injury or condition are rare. In these situations, paedophilic disorders have been discussed as behavioural manifestations of a 'pre-existing latent paedophilic urge' that has been 'released' by general impulse disinhibition. In other words, it could be considered that Gary's tumour did not cause a specific *change* in his sexual preference; rather, the preference had always been there, repressed or 'under the surface', and his tumour just allowed him to express it, as it made him disinhibited or unable to control his behaviour in general, including his sexual behaviour. This, in addition to his heightened sex drive, may have caused him to desire and indiscriminately search for any sexual partner, young or old, stranger or family member.

Others have argued that *de novo* (entirely new) sexual preferences are possible in the context of brain injury or disease even when general behaviour does not become disinhibited. Without more information, we cannot know if Gary had pre-existing paedophilia, but according to his own report, these desires only surfaced after his tumour developed. There were no previous complaints against him during his work as a schoolteacher. There is no suggestion that he was disinhibited in any other ways prior to his tumour removal. The fact that his paraphilia disappeared after his tumour was removed supports the notion of causation. Does that mean a change in sexual preference can arise anew and be switched off and on in anyone who has a scar, tumour or injury affecting a specific part of the sexual neural network? No, it doesn't – not all people with right orbitofrontal tumours develop paedophilia. Do all paedophiles have some kind of brain abnormality? I asked an expert in paedophilia this very question, and it turns out there's no simple answer.

DR TILL AMELUNG IS A GERMAN PSYCHIATRIST WHO HAS spent nearly a decade researching paedophilia. He became involved in a treatment program for people with paedophilia after he saw an interview with a man who described an intense feeling of 'falling in love' with a child. Ever since then, in Till's words, 'the topic has not let me off the hook'.

Germany is the only place in the world where there is a treatment program for paedophilia outside of the forensic system, Till tells me – in other words, it is a program for 'non-offending' people who have a sexual interest in children but have not committed any sexual offence. But is it actually possible to 'treat' paedophilia? Experts in the field disagree on exactly what should be treated. Some argue that the aim of treatment, which involves both psychotherapy and drug treatments, should be to alter and indeed eliminate the sexual interest in children. Others argue that this is actually not possible, as it is a sexual orientation that cannot be changed; rather, they say, the aim of treatment should be controlling the sexual impulses to prevent offending. It is a debate of 'elimination' versus 'control'.

Till is in the 'control' camp. After his treatment program, he says, those who have a sexual interest in children but have not offended report that, as a result of the treatment, their sexual interest and impulses have decreased. However there has not yet been a long-term follow-up study, so it is unclear how long the positive effects of treatment may last, and whether any of these non-offenders have ended up committing a child sexual offence. Till made a comment that I found both confronting and fascinating, and I have thought about it repeatedly since we spoke: 'There is a struggle of people with paedophilia to have a sex life without doing harm,' he said.

Most people with right temporal and/or frontal injuries do not develop hypersexuality or paedophilia. Those who do develop paedophilia in the context of a neurological condition typically show general behavioural 'disinhibition'. But why do only *some* people with brain disorders express this in a sexual way? What is it about Gary that is different from other people who have right frontal tumours and do *not* develop such dramatic sexual changes? Till suggested that people like Gary may have had a sexual interest in children before their injury or the onset of their neurological disorder. They may have a predisposition of some sort, which there is still very little understanding of, and this makes them more vulnerable to develop it after an injury or the onset of disease. It might be that the disinhibition that can occur after a brain injury, disease or tumour, as in the case of Gary, means that this level of response shifts. The latent response that only existed very subtly before the tumour grew may then suddenly turn into a full-blown sexual reaction.

There are multiple pathways to developing paedophilia. Prenatal and developmental predispositions have been identified, such as the masculinisation of the brain in utero, and some evidence of peer alienation during early puberty – that is, separation from same-sex peers due to trauma or some other type of atypical feature, such as a physical or psychological difference. So, are the brains of paedophiles somehow different? There is no simple answer, as much of the research is confounded by the fact that studies are conducted with those who have already committed a sexual offence and are in prison. In this population, one common finding is reduced amygdala volume. This is fascinating given that we know that when both amygdalae are destroyed or damaged in animals or people they can experience Klüver-Bucy syndrome, and considering my own research finding of a bigger amygdala resulting in a better sexual outcome after seizure surgery (see Chapter 1). Yet again it highlights that there is no doubt that the amygdala is fundamental to our sex drives and lives.

What about those people who have paedophilic interest but have not committed any offence? It turns out that their brains are actually different from the brains of those people who have offended. Those who have offended typically have reduced brain volumes, while those who have paedophilic interest but have not offended have larger brain structure volumes, higher connectivity and 'overcompensatory' mechanisms – that is, stronger 'brakes' when doing cognitive tasks that assess control. Till explained that 'normal' people's brains seem to be somewhere in between paedophilic offenders and non-offenders in terms of these specific measures – volume/size, connectivity and strength of 'brakes' during response control tasks. But this is all preliminary research that is still being carefully analysed.

Surely, though, it is possible that non-offenders could move into the offender group at any time; all offenders began their lives as non-offenders. So, if there are clear differences in brain structure and function between non-offenders and offenders, what causes this change? Do these brain changes occur after they have offended or just before? Research is currently underway to try to answer these questions and understand what predisposes non-offenders to offend. This is crucial research, as identifying these people and what triggers their offending is the first step in developing treatments to prevent it. Preventing offending is what is required to avoid the tragic and often life-destroying consequences of sexual abuse for victims.

I asked Till what reactions he has had from people when he tells them about his work. He said he is very cautious about discussing his area of research. Some people have told him about their own sexual abuse experiences, and he has even bumped into some of his patients by chance outside of work, so he feels that being careful with his job description is important for their privacy. What motivates him to continue working in the field? He told me that his main motivation was to address the discrepancy between the 'disastrous consequences' of paedophilia for everyone concerned, most particularly the children who are sexually abused, and the scarcity of knowledge about the causes of paedophilia and how to address them:

> The debate on whether to treat the sexual interest in itself or rather the self-control mechanisms is a good example of how even the most renowned specialists do not agree on the most basic question. From my point of view, much of this confusion is caused by concepts

that are not yet well defined, like, for example, mistaking 'paedophilia' as 'child sexual abuse' or the debate over paedophilia as a 'sexual orientation'. This imprecision in concepts results in imprecise research and unstable results. I feel that is a major problem also for clinical work. This lack of knowledge keeps patients and victims at a risk that might be reduced with better research. That's what I am hoping to be able to contribute to.

I understood why he had stayed in this field. There were so many unanswered questions to explore. So many lives could potentially be saved from the trauma of sexual abuse if we understood more about this condition, and how to identify and treat it before sexual offences occur.

WHILE GARY'S TUMOUR DISRUPTED HIS RIGHT FRONTAL lobe, another case that involved the right temporal lobe is frighteningly similar. Todd was 19 years old when he started to have frequent feelings of déjà vu – an overwhelming feeling of familiarity, as though he was reliving a moment he had already experienced – up to 20 times a day. This can occur when a seizure focus occurs in the hippocampus, which is a crucial memory structure (see Chapter 3). Over time, he developed other symptoms during his déjà vu attacks, including breathlessness, sharp chest pains and even occasional musical hallucinations, when he could hear a particular song 'as if it was playing in the next room'.

During some seizures he would also have gustatory (taste) and olfactory (smell) hallucinations. Despite trying various anti-seizure medications, his seizures were never well controlled, but he was still able to work in his role as a pharmaceutical researcher and also volunteered with his local epilepsy foundation. He had personality changes over time, including feeling more spiritual, creative and musical. He was well-liked by his work colleagues and loved by his wife, who had said he was a 'good husband', was 'not given to any inordinate or perverse sexual practices', and was always 'sensitive to her needs and feelings'.

Todd was 33 years old when he had his first neurosurgical operation. The seizure detectives, in the form of an MRI brain scan and video EEG, had found the culprit – his right temporal lobe. He had a right temporal lobectomy and afterwards had nine months of seizure freedom. He decided to reduce his anti-seizure medications, but this triggered a seizure, which was followed by many more. His seizures were different after his surgery, and started with a ringing sound, a foul smell and a feeling of thickness in his tongue. Six years after his first surgery he had a second operation – a posterior right temporal resection – to get rid of the remaining bits of his right temporal lobe that were still causing seizures.

The surgery was considered a success in one way, as he had four years with no seizures, but the dramatic changes in his behaviour that started a month after the operation were far from desirable: irritability, hyperphagia (excessive eating), hypersexuality and coprophilia (an interest in faeces and defecation). His neurologists diagnosed Klüver-Bucy syndrome. His hypersexuality manifested itself with his wife and when he was alone. In his own words, as reported in his case study by the famous Oliver Sacks and co-authors Julie and Orrin Devinsky, Todd said:

> My appetite for food and sex increased dramatically
> … I wanted sex constantly. Every day. I was very
> easily stimulated and began to touch myself regularly
> [and] request sex daily from my wife. If I wasn't
> having sex with my wife, I masturbated. This behavior
> increased over time. I became more emotionally
> labile, obsessive–compulsive, but on the other extreme
> disinterested or unable to initiate things I needed or
> was supposed to do. I raged for hours at inappropriate
> things at home … I become distracted so easily that I
> can't get anything started or done.

He started to watch pornography online while his wife slept. He became lost in a labyrinth of websites and was tempted to view and purchase child pornography. He became obsessed with it, and went on to download and purchase images of prepubescent girls. He was ashamed and secretive about it, and never discussed it with his wife or anyone else. Eventually he was arrested by US federal authorities for downloading child pornography.

Todd's wife told his doctors, 'He wanted to have sex all the time. He went from being a very compassionate and warm partner to just going through the motions. He didn't remember having just been intimate. He said it didn't seem like a current memory or emotion.' Until the day the federal officers showed up, she said, Todd had wanted sex five to six times a day, constantly touching and grasping at her. Two years after his arrest, he was prescribed antidepressive and antipsychotic medications that made him 'much warmer and loving' and stopped his angry outbursts, his wife said, though they left him with no libido at all. 'It is as if a faulty switch has been turned off,' his wife observed.

Todd's case is so extraordinary that Oliver Sacks writes about it in his posthumously published book, *Everything In Its Place* (2019), where he calls him Walter. We learn that his wife stood by him during his imprisonment and subsequent home confinement, and when he was freed they returned to their previous lives. Sacks writes that, during their last meeting, 'Walter' was 'relieved that he had no more secrets to hide. He radiated an ease I had never seen in him before'. (See also Chapter 10.)

GARY'S AND TODD'S BRAIN LESIONS WERE FOCAL, impacting on specific brain regions known to be part of the sexual neural network. Their neurosurgical operations targeted these areas and seem to have switched on and off the hypersexual and paedophilic behaviours. Other neurological conditions cause more widespread lesions or damage, and if multiple lesions invade many different parts of the sexual neural network, they can cause havoc with sexual behaviour. Multiple sclerosis (MS), which means 'many scars', is a neurological disorder characterised by lesions that can occur throughout the central nervous system – in the brain, spinal cord and optic nerves (the nerves that connect the retinas in the back of the eyes to the brain). It is an inflammatory disorder in which white blood cells, which usually fight infection, cause injury in the nervous system. The myelin sheath which protects nerves is damaged during the inflammatory process and this means the nerves cannot conduct electricity effectively. The lesions typically occur in the brain's white matter – the myelin of nerve cells. MS symptoms depend on where the lesions are, and can range from tingling toes and blurred vision to incontinence and slurred speech. In rare cases, like Debbie's, a multitude of sexual changes can occur due to lesions throughout the sexual neural network.

Debbie was 26 years old when she first felt numbness in her hands and legs. After it persisted for a few months, she decided it was time to see a doctor. She had a neurological review and a lumbar puncture to check her cerebrospinal fluid, and was diagnosed with MS. She was given a hormone treatment that was very effective and the feeling in her hands and legs returned. She carried on with her work as a waitress and caring for her two boisterous sons. Even though she had no symptoms and the diagnosis felt like it might have been a dream, it still popped into her mind occasionally and kept her awake at night; she wondered if and when the brain lesions would strike again and what her future held. She knew it had struck again a year later when she had an intermittent feeling of pins and needles in her legs, feet and left hand. She tried to ignore it, but when her vision got blurry and her left leg felt like it was dragging while she walked, it was time to go back to hospital. Her CT brain scan was normal and she was given the same treatment as the previous time, but it didn't work as well. As a single parent she had to carry on and return to work, but she was exhausted after her waitressing shifts.

Debbie's next recorded hospital visit was a couple of years later when her sensory and motor symptoms had switched to her right side. Then there were no hospital notes for the next nine years, until 1991, when she was 39 years old, and was arrested for a variety of sexual offences. The case report, by psychiatrist Neil Ortego and colleagues, outlines that over a two-month period Debbie was alleged to have

> seduced her 14-year-old son's 17-year-old female lover; engaged in oral sex and used a vibrator during mutual masturbation with the 17-year-old girl; offered sex to her son's 15-year-old male friend; requested that her son and his girlfriend engage in sexual intercourse while she watched; attempted to have the pet German shepherd have intercourse with her in front of her son and his girlfriend; had her pet dog lick her genitalia in front of minors.

Debbie had been heterosexual, with no prior history of any paraphilia; she had not even been sexually active in the years preceding these criminal offences. Yet here she was, 13 years after her diagnosis of MS, with a sudden onset of a myriad dramatic sexual changes: hypersexuality, incest, zoophilia (engaging in sex with animals), scoptophilia (arousal from watching others have sex), ephebophilia (sexual activities with adolescents), paedophilic disorder and exhibitionism. These changes in her sexual orientation and behaviour were prominent symptoms, and they did not appear to be caused by a general behavioural disinhibition.

The case study reported that she was seen by a psychologist during her legal proceedings. Her assessment could not be completed due to her 'slowness and limited cooperation'. The psychologist found her to be 'demanding and narcissistic', and noted she was depressed. The question of whether her MS had altered her impulse control was raised by the psychologist in her report, but it was never discussed in her criminal trial; the case report doesn't say why. The finding that Debbie essentially had 'faulty brakes' due to the extensive brain damage caused by her MS raises questions about her responsibility for her actions – and whether she deserved to be punished. This is discussed further in Chapter 10.

During her trial, Debbie woke one morning and was unable to get out of bed, and had lost control of her bladder and bowels. She went to the emergency department and was hospitalised, but it was suspected that she was malingering – feigning illness to avoid the repercussions of the trial. Nevertheless she was sent for an MRI brain scan, which revealed that lesions consistent with MS were scattered throughout her brain, and – crucially – were found in multiple critical areas of the sexual neural network. It showed extensive white matter lesions in her frontal, temporal and parietal regions, and a large lesion in her left thalamus and right midbrain. Essentially, there was widespread damage throughout her brain. After treatment with high-dose steroids she was able to walk with the aid of a walking frame, but the steroid treatment was discontinued after she was discharged from hospital.

Back in the courtroom, the prosecutors in Debbie's case alleged that she had molested her older son 12 years earlier, suggesting that she was a paedophile even before the MS lesions had ravaged her brain. This was never proven; the nature of this alleged earlier incident – planned, organised and done in secret – was entirely different from her dramatically altered and exhibitionist sexual behaviour in the two months before her arrest. Without further information it is impossible to know if Debbie had a *de novo* onset of her paraphilias, or had acted on long-standing but repressed paraphilic preferences after the onset of her MS.

Despite the clear evidence of MS, Debbie was convicted on all counts of child molestation and given a maximum sentence of eight years in prison. The final months of her life were spent lying in her jail cell, 'incontinent of urine and faeces'. The prison guards thought she was malingering so she was left unaided, with a mop to clean up after herself. She was described as being 'surprisingly unconcerned by her predicament'. She died unexpectedly – and perhaps mercifully, given the tragedy of her situation – of a pulmonary embolism, a blood clot in the artery that runs from the heart to the lungs.

Despite the clear association between their neurological conditions and paraphilic disorders, Gary, Todd and Debbie were all charged and convicted of sexual offences, and two of them spent time in jail. These cases present a complex challenge for the criminal justice system, which we will discuss in Chapter 10.

The cases I have included in this chapter are rare. The vast majority of people with temporal lobe epilepsy or right frontal lobe tumours do not develop paraphilias like Harry, Gary and Todd did; very rarely do people with MS develop paraphilias like Debbie. But these intriguing cases offer a chance to gain insights into how our brains control our sex lives. We need to keep asking *why* and *how*, especially in the case of paedophilia, where the consequences for the person and their potential victims are so devastating. It is only by clarifying the nature of this condition, identifying its causes and understanding how it can lead to criminal behaviour that we will be able to prevent the immense suffering it brings.

8

SEX ON THE SPECTRUM

When I was a student I had a job as a support worker with my university's disability liaison unit, a service that assisted students with disabilities. I worked with several students, including a man with cerebral palsy who was studying computer science, and a woman who was blind and completing an arts degree, both of whom needed a note taker in their lectures. I also sat with a woman with Tourette syndrome during her exams, which she completed in solitude so that her frequent verbal and motor tics (singing snippets of radio ads and pop songs, clapping her hands and randomly jumping up out of her seat) would not disturb other students. The most memorable student I worked with was George, a man the same age as me, who had a diagnosis of Asperger syndrome – what many refer to now as 'high-functioning autism spectrum disorder'.

Asperger syndrome and autism were first described in the early 1940s by two Austrian physicians, Hans Asperger and Leo Kanner respectively. Coincidentally they both used the term 'autistic' to describe children, typically boys, who had difficulty interacting socially and had repetitive, restricted interests. These children also showed a range of language and cognitive abilities, from complete absence of speech and intellectual disability, to highly developed skills in specific cognitive domains. The symptoms of Asperger syndrome as characterised in the original description were social difficulties, such as isolation or lack of reciprocity in social interactions; normal to advanced language acquisition but subtle abnormalities in verbal and non-verbal communication, such as pedantic vocabulary; a lack of variation in the pitch of their speech; a narrow and restricted focus of interests; advanced cognitive skills in specific areas; and motor clumsiness.

Asperger syndrome didn't receive much attention until about 50 years after Hans Asperger's original research was published, when his work was translated into English by world-renowned autism researcher Uta Frith. Another famous autism researcher, Lorna Wing, had proposed in the 1980s that Asperger syndrome and Kanner's autism were part of an 'autism spectrum'. Nevertheless, the idea that they were two separate disorders had already been unleashed, and garnered much scientific attention and public interest from the 1990s onwards. Many authors distinguished the conditions on the basis that children with Asperger syndrome had normal development in their first few years of life, and their cognitive and linguistic skills were normal or enhanced. Furthermore, some argued that there were qualitative differences in the nature of the social impairment in that children with autism were completely uninterested in other people, while those with Asperger syndrome tried to engage with others but did so in a dysfunctional way.

Unlike autism, which remains in the current version of the *Diagnostic and Statistical Manual of Mental Disorders* (DSM-5) as 'Autism Spectrum Disorder', Asperger syndrome only had a brief existence as a DSM diagnosis in the fourth version (DSM-IV), after which it was dropped. The condition was embroiled in controversy from the outset, mainly because of the diagnostic criteria. If someone showed significant impairment in social interactions and restricted interests, then they met the criteria for autism – meaning that they could not then be diagnosed with Asperger syndrome. Ironically, even Hans Asperger's original cases would have failed to qualify for a diagnosis of Asperger syndrome according to the DSM-IV. It was also unclear whether Asperger syndrome and 'high-functioning autism spectrum disorder' could actually be distinguished from one another through clinical assessment or neuroimaging. Research showed only subtle differences, if any, and eventually it was decided that there was insufficient evidence to warrant a distinction between the two diagnoses. This led to the two conditions being merged under the umbrella term of 'autism spectrum disorder' (ASD), with Asperger syndrome now typically referred to as high-functioning ASD.

There is still debate about whether Asperger's and high-functioning ASD are actually the same thing. Many people who had been diagnosed with 'Asperger syndrome' were shocked when their diagnosis suddenly ceased to exist. It was considered an 'identity-defining diagnosis'; many had adopted the term 'Aspy'. The 'neurodiversity' movement argues that diagnostic labels of autism or Asperger syndrome should be embraced and accepted as normal variations of humanity, each with unique benefits that are worthy of being celebrated and utilised by society. This topic is eloquently addressed by Steve Silberman in his book *NeuroTribes*, in which he writes:

> Autistic people have always been part of the human community, though they have often been relegated to the margins of society … Whatever autism is, it is not a unique product of modern civilization. It is a strange gift from our deep past, passed down through millions of years of evolution. 'Neurodiversity' advocates propose that instead of viewing this gift as an error of nature – a puzzle to be solved and eliminated with techniques like prenatal testing and selective abortion – society should regard it as a valuable part of humanity's genetic legacy while ameliorating the aspects of autism that can be profoundly disabling without adequate forms of support. They suggest that, instead of investing millions of dollars a year to uncover the causes of autism in the future, we should be helping autistic people and their families live happier, healthier, more productive, and more secure lives in the present.

The controversy surrounding Asperger syndrome continues, with claims that Hans Asperger was an active Nazi collaborator and participated in a child euthanasia program during World War II coming to light in early 2018. There have been calls for his name to be 'scrapped from the medical lexicon', with many previous supporters of the term 'Asperger syndrome' abandoning it.

My first introduction to Asperger syndrome was when I met George. He was well over six feet tall and obese – or, as he prefers to say, he has a 'large mass'. When I was his support worker he wore his pants pulled up high over his belly, and he tended to bounce rather than walk – there was no missing him as he bobbed around the university. He would talk to himself, or approach strangers and shout out what was on his mind. When I first met him he was doing a science degree and needed support to make sure he went to his lectures, to keep him on track in classes, and to try to prevent him from shouting out and disrupting his fellow students. His special interests were the weather, the environment, train timetables, cricket statistics and Jeff Kennett, the then-premier of Victoria, who George particularly disliked. One of his favourite phrases was: 'Who's bad? Jeff Kennett!' He once shouted this in the microphone in front of a lecture hall full of students and proceeded to launch into a lengthy rant about all that was bad about Jeff. The students stared and laughed, and I had to physically pull George away from the microphone and into a seat when the stunned lecturer finally arrived.

George's knowledge of the local train timetables was astounding. If I told him I was catching a train home to visit my parents, he would provide detailed advice about my options. 'You absolutely must get the 534 as this is an express,' he would say. 'The 614 will stop all stations and take you 27 minutes longer. Alternatively, you could catch the 555 from Spencer Street and that will take you through the loop and then only stop at three stations ...'

When we walked between lecture theatres on the university campus, George liked to link arms with me. I never felt self-conscious about this. Rather, I was chuffed that someone who had known difficulties connecting socially with people seemed to have a connection with me. It was his way of showing he was comfortable, and I never felt anything other than a sense of camaraderie. I think he liked being as close as possible to my ear, to ensure I was listening to his views about whichever of his pet topics was engulfing his brain at that moment. In winter, he often walked around with his hands down his pants. This typically resulted in concerned or puzzled looks from people passing by. I recall saying to him several times, 'George, you really need to take your hands out of your pants right now. It's best not to have them in there while you're walking around in public.' His response each time was: 'But it feels so warm in there.'

He had no idea of the social inappropriateness of this simple act of keeping his hands warm, or of the possibility that others could interpret this as a potentially sexual act. It was purely a sensory thing to him. I can't recall if he ever tried to link arms with me with one hand down his pants, but even if he had, there certainly would not have been any sexual element to it. There was, however, no denying the sexual nature of public masturbation, which George engaged in several times.

He was first caught in the university library, one of his favourite spots, and on another occasion on a couch outside the faculty office. This was no doubt a very stressful situation for all those involved: the witness, the security guard called to deal with George, and of course George himself, who due to the nature of his condition had difficulty comprehending what all the fuss was about. On one occasion, when we were about to enter a lecture hall, George suddenly announced that he had to go to the toilet. I was waiting for over ten minutes and thought I'd better check on him. I walked down the hall towards the men's toilets, and as I approached the toilet door I could hear him talking to himself about 'astronomy and astrology' while making sounds of pleasure. I opened the door just a crack and said, loudly, 'George, your lecture started ten minutes ago. We really need to go now.'

'Yes, I'm coming, I'm coming.'

I shook my head and smiled at the pun.

On another occasion, I was waiting outside a lecture hall for George to arrive when a student who recognised me as his support worker grabbed my arm, looking frantic. 'Are you waiting for George? Can you please go to the faculty office? They don't know what to do. He's having a major meltdown.' She didn't even wait for my response but turned and ran back down the hall. I felt my stomach clench as I followed close behind her, wondering what state he would be in, and what they expected me to do.

I heard him long before I saw him, his penetrating voice booming down the hall. He was sitting behind a desk, intermittently slamming the phone receiver into his chest and shouting into it. There were at least six university staff members backed against the office wall, all staring at him, their facial expressions a mix of fear and fascination. I went straight to George and put my hand on his shoulder. 'George, it's Amee. You need to stop shouting so we can talk. Come on, we have a lecture to go to.' He looked at me, took a deep breath and dropped the phone on the desk. I kept my hand on his shoulder as he stood up, and we walked out of the office. I saw one of the staff raise their eyebrows and whisper, 'Wow.'

I can't recall the incident that triggered this outburst but heard later that he had been shouting for over 15 minutes and no one had known what to do. I hadn't known what to do either, but my advantage was that George was familiar with me. I think that my instinct to touch him was the key to ending his rage. He could look frightening – his mere size was daunting enough, let alone when it was combined with shouting and slamming a phone into his chest. Reflecting on these incidents, I find it interesting that no one had any qualms about approaching him while he masturbated in public. I guess sex is more predictable than rage.

Is George's behaviour, in particular his public masturbation, typical for someone on the autism spectrum? I recall his mother saying to me years ago how she felt sad that he would probably never have a sexual relationship with anyone other than himself. How do the symptoms of ASD, in particular impaired social and communication skills, impact on sexual interactions, which could be considered some of the most complex and demanding of all human social behaviours? Are the stereotypical beliefs that people who are 'on the spectrum' are asexual and uninterested in romantic relationships actually true?

THE RESULTS OF RESEARCH INTO THE SEX LIVES OF people on the autism spectrum is mixed. Some studies have found differences in sexual knowledge, experiences, attitudes and behaviours between people with ASD and 'neurotypicals' (those without the condition), while others have found no differences at all. This inconsistency is not surprising given the heterogeneous nature of ASD, and the methodological variations between studies. One recent review of the issue shows that people with ASD do seek out sexual relationships, and experience the full range of human sexual activities and behaviours. In some cases, however, the core symptoms of the condition, combined with limited sexual knowledge and experience, can lead to the development of challenging sexual behaviours, such as hypersexuality, paraphilic disorders (see Chapter 7) and, in rare cases, even sexual offending.

Hypersexuality, and paraphilic fantasies and behaviours, have been found to be more common in adult males with ASD than in neurotypical people; some researchers have hypothesised that the restricted interests and repetitive behaviours commonly seen in ASD may transform into sexualised behaviours in adulthood. Furthermore, a sensitivity to sensory stimuli, in particular touch, can lead to an overreaction or underreaction to sexual stimulation. A caress that may feel deliciously sexy to a neurotypical person could be experienced as a commonplace interaction, like a handshake, to someone on the spectrum.

Excessive masturbation is common in males who have ASD. For example, in one case study it was reported that a 17-year-old man with a diagnosis of Asperger syndrome had intrusive sexual thoughts and urges, and was masturbating 25 to 30 times a day, resulting in penile ulcers. It is unclear whether this excessive behaviour represented a repetitive behaviour, was part of an obsessive-compulsive disorder (common in people with ASD), was the result of sensory hyposensitivity, or was the only means of him being sexually active due to limited social skills. Public masturbation was the most common type of inappropriate sexual behaviour reported by carers or parents of adolescents and adults with ASD. So it seems that George's exploits in the university library were not unusual for someone on the spectrum.

Why are these sexual behaviours found more frequently in males with ASD? First, there are more males than females diagnosed as being on the spectrum, at a ratio of somewhere around 3:1 or 4:1, and much of the research about sex on the spectrum has focused on men. Interestingly, a study published in 2017 by Daniel Schöttle and colleagues that directly compared hypersexuality in people with ASD and neurotypicals found that there were no differences between females on the spectrum and those who were neurotypical. The authors of this study argued that this gender difference was due to the observation that females with ASD 'seem to be better socially adapted and usually show less pronounced ASD symptomatology'. But did the study find that rates of hypersexuality in males with ASD were higher than in the general population? Yes, it did. Previous studies of the prevalence of hypersexuality in healthy males give estimates of 3–12 per cent, and in the males with ASD in the study, the prevalence of hypersexuality was 30 per cent.

This same study also looked at rates of paraphilias in people with ASD. The most common paraphilias reported by males with ASD were voyeuristic fantasies (39 per cent) and behaviours (28 per cent), while in the neurotypical group such fantasies and behaviours were reported in 10 per cent and 3 per cent of subjects respectively. *Masochistic* (sexual arousal from suffering physical pain or humiliation) and *sadistic* (sexual arousal from inflicting physical or psychological pain on a sexual partner) fantasies and behaviours were also reported more often by males with ASD than those without. It is thought that this may be due to a high level of sensory hyposensitivity in people with ASD – that is, lack of sensitivity – and the corresponding need for higher levels of stimulation to become sexually aroused. Furthermore, a heightened sex drive may mean that people with ASD are more likely to act out their sexual interests and to seek novel sexual activities. When comparing paraphilias in females with and without ASD, the only difference the authors found was that masochistic behaviours were reported more frequently by females without ASD. So once again, there is a gender difference here. As with hypersexuality, these rates are also higher than prevalence estimates found in general population studies.

When it comes to these novel sexual activities, I was surprised to find an example of the rare paraphilia called zoophilia in a 17-year-old with ASD, who I'll call Sam. He lived in rural Sri Lanka. As a toddler, he had shown some classic ASD symptoms – playing alone at preschool, not sustaining eye contact and not engaging in imaginative play. He had demonstrated restricted repetitive behaviours such as lining up toys, delayed echolalia (echoing other people's speech) and an extreme need to only eat particular foods. His speech had regressed when he was two years old until he spoke only single words. He was diagnosed with ASD at five years of age and treated with behavioural interventions. After a few years his parents stopped taking him for reviews due to the travel costs. Over the years his speech improved, as did his restricted and repetitive behaviours, but he still had difficulty establishing peer relationships.

In the year before the authors of the study met him, his parents noticed Sam was spending more time at his grandparents' cowshed. They didn't intervene as they thought he was developing an affinity or 'warmth' for the animals, which was a rare thing for Sam to show. A neighbour saw Sam masturbating near the cowshed one day and told his mother. Sam's father warned him not to engage in this behaviour publicly but did not think of any connection to the cattle. 'Things took a twist when his grandmother found him having penetrative intercourse with the heifer. She had screamed and reportedly had fainted,' the authors of the study noted.

Sam's parents immediately took him to a local healer in their village. He was prescribed a 'thovil' – an exorcism ritual – to relieve him of 'demonic possessions', but Sam's parents could not afford the ritual and instead took him to the free local health service the next day. He was referred to the youth psychiatric service and, after initially denying his sexual urges, he eventually requested help. He reported that for the preceding year he had experienced recurrent and intense sexual fantasies about having sex with cattle, and had masturbated while observing them and watching them mate. Acting on these urges had meant he had missed school. He had never had a romantic or sexual relationship with another human; although he had watched (human) internet pornography, he said that he did not enjoy it compared to watching animals. He denied any sexual interest in non-living objects, children or adult males, but acknowledged some sexual interest in adult females.

Sam was treated with the antidepressant Sertraline, cognitive behavioural therapy and 'orgasmic reconditoning' that aimed to reduce inappropriate sexual arousal. (Orgasmic reconditioning is a controversial treatment that has little empirical support; it involves a person with 'deviant' sexual preferences masturbating to 'deviant' materials, which are then switched to 'non-deviant' materials just prior to orgasm – and earlier in the process as the treatment progresses. The idea is that eventually the person's sexual pleasure will be diverted to the non-deviant material.) Sam was also given sex education and social skills training. After three months of treatment, Sam reported better control of his sexual urges and reduced psychological distress. His parents had not observed any further deviant sexual behaviour. It is important to note that this is the only reported case of zoophilia in a person with ASD. Sam's case is an extreme example of how ASD-associated difficulties in establishing human sexual relationships may be associated with the development of a paraphilic disorder.

Is there anything different about the way the sexual neural network operates for people with ASD that might help explain the higher rates of some sexual behaviours? ASD is neurodevelopmental in origin, meaning that the brain changes start in utero and continue over time. It's also important to remember that there are enormous variations in how ASD affects people, so understanding and describing the neurobiological mechanisms underlying ASD is very difficult. Nevertheless, based on numerous neuroimaging studies that have compared the brains of people with ASD and neurotypicals, there is a consensus that neuroanatomical structure, functioning and connectivity do differ between the two groups. It is far beyond the scope of this chapter to canvass this research thoroughly, but one recent study did turn up a few interesting points about specific brain regions that are relevant to us here.

In the largest neuroimaging study to date of people with ASD – 1500 people were studied, ranging in age from two to 64 years – a group of international experts found that compared with neurotypicals, people with ASD had a larger total brain volume and cortical thickness, and smaller volumes of specific brain structures including the amygdala and parts of the basal ganglia (nucleus accumbens, putamen). The ASD group also showed increased thickness of the frontal cortex, and decreased thickness of the temporal cortex. The basal ganglia, frontal and temporal brain regions all play a role in the sexual neural network. From a broader perspective, these brain regions are also part of what has been termed the 'social and emotional brain', a larger network of brain structures that are involved in processing social and emotional information, such as empathy, face processing and self-referential processing.

As psychiatry professor Christine Ecker highlights in her overview of these neuroimaging findings, it is important to note that 'the neuroanatomical differences are neither unique to, nor causal for ASD'. In regard to the differences in sexual behaviour between people with ASD and neurotypicals, there has been no study to date that has investigated whether differences exist in the brains of those with ASD who have paraphilias or hypersexuality as compared to those with ASD who do not have these specific sexual behaviours. So although we know that, overall, it appears that people with ASD do have brain differences in regions that are part of the sexual neural network, not all people with ASD experience sexual changes, and we don't know if there is something unique about the brains of those people who do. This is a question for future research.

WHEN PEOPLE WITH ASD DO FORM SEXUAL RELATION-ships, it has been reported that rates of homosexual or bisexual orientation are higher than in the neurotypical population. There have been various proposals for why this might be, including that people with ASD may be less sensitive to gender roles or social norms, or that gender is less relevant to them when searching for or finding partners. I don't know if George ever had a girlfriend or boyfriend, or any friends at all, nor do I know any more about his sex life than what I knew about him during my time as his support worker. I do feel that we were – and still are – friends; we have maintained contact since I finished working for the disability liaison unit decades ago.

George successfully completed his Bachelor of Science, followed by a Master's degree, and also commenced a librarianship course at a different university. Unfortunately, he got into a scuffle with a security guard and was told he was never allowed back on campus, so he was not able to finish this course. I was enraged when I heard about this, and presented his case at an international autism conference with the title 'University challenge: A student with Asperger Syndrome shares his rage'. He would have made a wonderful librarian or transport information officer. It would take an understanding employer to take him on, but the benefits of utilising George's unique gifts would surely be worth it.

George used to call my mobile phone at random times. There was never any greeting, just a sudden announcement about his current 'mass', or his views on the weather or the latest government environmental policy. Over the years he called less frequently, and now he prefers to write letters. A few times a year a computer-printed, hand-signed letter will arrive; I recognise his handwriting on the envelope immediately, and always enjoy reading about his life. I laughed when reading the most recent one, as I could hear his distinct voice using the very same words decades ago, when he would question me about my preferred weather and request an exact temperature in degrees Celsius. In this letter, he wrote about a

> dreadful 38 degree Friday earlier this month. That day was some of the worst weather I have been out in, with dust and debris making the awful heat even worse – and Mummy complained that I still accosted people about whether they would prefer it to be 15 degrees instead (most agree but say 25 degrees is what they are looking for, as I have known since 2001).

Each time a letter arrives, I reflect on the time I spent working with George when we were both students, and my reflections always make me smile. He has given me insights into neurodiversity, and what sex might be like for someone on the spectrum, that I have never forgotten.

9

PORN ON THE TRAIN (AND ON THE BRAIN)

I had a surprising experience on a peak-hour train one morning. I was sitting directly behind a man who had his mobile phone held up at eye level, and I could see the moving image of what he was watching reflected in the window. I couldn't figure out what it was for a while, then suddenly recognised a graphic close-up of genitals in full swing! He was watching pornography, on his morning commute, completely oblivious to the passengers around him. I was more curious than offended, and watched the reflection in amazement, wondering how long he would persist. Then I started wondering about what was going on in his brain. Was he born with an appetite for porn, or did his apparently prolific and blatant consumption change his brain to want or need more of it?

As I contemplated these questions and watched the reflection bouncing around on the window, the train carriage started filling up with other passengers. A woman sat next to me, and it wasn't long before she noticed what our fellow commuter was up to. She gasped and we exchanged wide-eyed glances. She commented that her partner was a police officer and that she should call him, then sighed and muttered, 'Disgusting.' Suddenly she leaned forward over the top of his seat and slapped his mobile phone down, telling him, 'You're a nut short of a fruitcake, mate!' He tucked his phone away and sat quietly until the train's next stop, when he scurried out of the carriage without a glance or a word. It was certainly one of the most interesting train trips I have ever experienced.

The internet has revolutionised the porn industry, and it has been estimated that 50 per cent of all internet traffic is related to sex. The three As that make internet porn such a mass phenomenon are its accessibility, affordability and anonymity. Are we really so blasé about porn now that people feel they can watch it openly on their daily commute? Was the train passenger who sat in front of me that day a typical porn viewer, or was he an anomaly, a 'high porn consumer' who had lost all inhibitions, his brain 'hijacked by porn'?

Is it really possible to become addicted to porn, and can it actually change our brains? These are highly controversial, challenging and complex topics. I want to preface this discussion by saying that as a clinical neuropsychologist, I don't have any experience in treating or even meeting people with so-called porn or sex addictions. I'm raising these topics here because I was curious after my 'porn on the train' experience to understand more about the issue, and what we know (or as I found out, don't know) about 'porn on the brain'. So here I'll focus on what some neuro-scientific studies on this topic actually show.

It turns out that the brains of people who are consid-ered 'high porn consumers' are different in both structure and function from the brains of those who watch porn only occasionally or not at all. The first study to explore this issue was published in 2014 by Simone Kühn and Jürgen Galli-nat; they scanned the brains of 64 men who had reported their hours of porn consumption. (The average was 4 hours a week, but consumption ranged from 0 to 19 hours a week.) The researchers found an association between the number of hours that pornography was viewed, and the size and activity of certain brain structures. Specifically, high porn consumers had smaller volumes of the right caudate, part of the striatum – a 'deep brain' structure that is part of the basal ganglia and is known to be involved in reward pro-cessing (as discussed in Chapter 6). High porn consumers also showed less activity in the left putamen (also part of the basal ganglia) when viewing explicit sexual images, and less functional connectivity between the right caudate and left dorsolateral prefrontal cortex (in the frontal lobe) when compared with low porn consumers.

The dorsolateral prefrontal area is interconnected with other parts of the prefrontal cortex that mediate an incred-ible array of social, emotional and cognitive functions. The authors consider the dorsolateral prefrontal cortex to be 'a key area for the integration of sensory information with behavioural intentions, rules and rewards'. The 'fronto-striatal' network (which refers to the connections between the frontal regions and basal ganglia) is thought to help us choose the most relevant motor action in response to stimuli. Dysfunctions in this network have been related to inappropriate behavioural choices, such as drug seeking, which occur regardless of negative outcomes. So the functions of the brain regions in which differences were found are relevant to porn viewing, an activity that involves reward and motor actions, and engages social and emotional processes.

Although there were clear differences between the brains of men who viewed a lot of porn and those who viewed a little, what these findings do *not* tell us is whether viewing pornography actually *caused* those changes. It could be that these brain differences are pre-existing in men who watch a lot of porn, and predispose them to this behaviour. In other words, having a smaller right caudate and a less active left putamen could make you desire porn more than other people, and mean that you need more external visual stimulation to experience pleasure; you would therefore find porn more rewarding than other people might, leading to higher levels of consumption. Rather than being a consequence of frequent pornography consumption, these brain differences might actually be a precondition for it. The research done so far cannot answer this question of what causes what; it only tells us that the brains of high porn consumers are different. Interestingly, the high porn consumers were more likely to be depressed and have an alcohol use disorder; both these conditions are known to cause brain changes, so this adds another layer of complexity that needs to be addressed by future research.

IT'S ALSO NOT POSSIBLE TO TALK ABOUT 'ADDICTION' to porn without raising a related proposed condition: 'sex addiction'. Most of us know the term due to several famous people who have been labelled 'sex addicts' or have claimed the title themselves, such as former US president John F Kennedy, actor Billy Bob Thornton and rapper Kanye West. In the TV show *Californication*, David Duchovny's character Hank Moody is considered a textbook case of a person with 'sex addiction'; in real life, Duchovny actually checked into a rehabilitation centre specifically for this reason. Despite the availability of specialised private clinics, rehabilitation centres and therapists offering treatment for 'sex addiction', it is a highly controversial condition and there is ongoing debate about whether it really is a bona fide addiction, and whether it should actually be formally classified as a mental or behavioural disorder.

I spoke with David Ley, a practising clinical psychologist, author and sexuality expert who has written several books including *The Myth of Sex Addiction* (2012). 'The field of porn addiction is a mess, littered with pseudoscience, moral biases, and conclusions drawn in advance of the data,' David said to me. Fundamentally, he said, the 'sex and porn addiction' proponents argue that 'there are universal, homogeneous effects' of sex and porn consumption on people, but 'this is far from true'. 'There's a wealth of data demonstrating that alleged effects of sex or porn are extremely individualised, sensitive to personal, social, cultural and environmental contexts,' he argues.

So how did the concept of 'porn addiction' arise in the

first place? David highlighted the cultural and religious factors that propelled the idea forward. Religious institutions in the 1980s first popularised the idea to promote and drive legislation against pornography access; this was also around the time of the AIDS crisis, when there was a lot of fear about sex causing death. He said conservative churches 'latched onto the concept of porn and sex addiction to outsource moral management to a pseudo–health care industry. If it's a disease and not a moral issue, the church doesn't have to manage it – they just tell the person to go and get treatment.'

The first book on sex addiction was written in the early 1980s by Patrick Carnes; in it Carnes identified homosexuality as a disease and addiction (though he did not repeat the claim in his subsequent books). Carnes applied to sex a 12-step model of addiction treatment usually used for drug and alcohol addictions. 'It's a nice theory but where's the data?' David queried, adding that there are no published, peer-reviewed, empirical studies that show that this treatment model is effective.

In the early 2010s, anti-pornography religious groups began to promote porn addiction as a brain disease, and the notion of porn changing the brain became a hot topic of discussion. But for David, the whole concept and industry of porn/sex addiction is based on conservative moral values around sexuality that intrude into clinical practice: 'These conditions have been rejected by scientific and medical groups for well over 40 years, but are clung to by dedicated folks who have deep moral issues around sex, or are financially dependent upon the concept.' He also pointed out that 'across most definitions of sex addiction, masturbation, promiscuity and forms of infidelity are consistently identified as symptoms, and within religious individuals, same-sex desires are perceived as "addictive"'.

I also found it fascinating when David described how the concepts of porn and sex 'addiction' were supported much more in the United States than anywhere else in the world: 'Other countries do not have the huge, lucrative, entrenched industry that we do, which is, notably, clustered in highly religious states.' He gave the example that the state of Utah has the highest number of conservative Mormons, the highest level of porn consumption, and the highest number of porn/sex addiction treatment centres. So what is it about the United States that is so special in regard to this phenomenon? David's view is that 'it emerges from the American conflict about sex. We are obsessed with it and afraid of it at the same time.' The high numbers of religious conservatives in the United States is another factor, he argued.

In David's experience, he said, many advocates for the sex or porn addiction concept are themselves self-identified 'sex addicts'. 'That raises serious concerns about the possible intrusion of confirmation bias,' he noted. Another issue he raised is that most mental health professionals have little or no specific training in human sexuality. He quoted Alfred Kinsey, the famous American sex researcher with an institute that bears his name, who said that the definition of a nymphomaniac – or 'sex addict', in today's language – is someone who's having more sex, or different sex, from what the therapist is having.

The notion of excessive sexual behaviour has been around for hundreds of years. American physician Benjamin Rush (1745–1813) described various mental disorders he believed arose from 'indulgence of the sexual appetite', while nineteenth-century sexologists Richard von Krafft-Ebing, Havelock Ellis and Magnus Hirschfeld all reported cases of both men and women who displayed excessive and 'maladaptive' sexual appetites and compulsive masturbation. In the twentieth century, the terms 'Don Juanism' or 'satyriasis' were adopted for males and 'nymphomania' for females who showed such behaviours. In the third edition of the *Diagnostic and Statistical Manual of Mental Disorders* (DSM-III-R) in 1987, sexual addiction was defined as 'distress about a pattern of repeated sexual conquests or other forms of nonparaphilic sexual addiction, involving a succession of people who exist only as things to be used'. Later versions of the DSM dropped the terms 'non-paraphilic' and 'sexual addiction' as there was not enough research and no agreement about whether sexual behaviours could constitute an addiction.

In addition to 'sex addiction', excessive sexual behaviour has also been labelled 'hypersexual disorder', 'problematic hypersexual behaviour' and 'compulsive sexual behaviours'. More recently, the term 'compulsive sexual behaviour disorder' has been adopted in the first formal recognition of this phenomenon as a diagnostic entity, appearing in the most recent and eleventh version of the World Health

Organization's International Classification of Disease (ICD-11) in late 2018. 'Compulsive sexual behaviour disorder' is listed as an 'impulse control' disorder rather than being grouped with substance abuse disorders and other addiction-type conditions. The World Health Organization took a cautious approach, acknowledging that 'we do not yet have definitive information on whether the processes involved in the development and maintenance of the disorder are equivalent to those observed in substance disorders, gambling and gaming'.

The concepts of 'porn addiction' and 'sex addiction' are not currently recognised by the major sex therapy organisations and associations in the United States; in fact, the American Association of Sexuality Educators, Counselors and Therapists (AASECT) put out an official statement saying that (1) it did not find 'sufficient empirical evidence to support the classification of sex addiction or porn addiction as a mental health disorder'; and (2) that it found that sex addiction 'training and treatment methods', and the theory behind them, was not 'adequately informed by accurate human sexuality knowledge'.

I asked David about the commonly expressed notion that porn use shared similarities with other types of addictions, like substance use. There's a stark difference in physical effects and the process of 'withdrawal', for one, David noted: 'No one in the history of the world has ever died from blue balls, or not getting to have sex when they want to. But removing alcohol from a long-term alcoholic can and does result in seizure and death.' What about similarities in relation to porn use for things such as 'habituation' – the brain becoming less responsive to a stimulus over time, and needing more and more of it to feel the same reward – and the related notion of 'escalation' (the progression from 'vanilla' porn to more extreme porn)? 'There is no evidence that porn results in a habituation or desensitisation effect that is distinguishable from normal changes in sexual development,' David said to me. 'A great deal of research on porn consumption actually demonstrates that people on average watch the same kind of porn over and over, and there is no evidence of a slippery slope leading to more "extreme" types of porn over time.'

David said that in most studies on porn use, 'extreme' use is typically categorised as only 17 minutes to one hour per day, while the 'average' use is 10–15 minutes a week. 'Is it *really* changing brains at that level of use?' David asked, 'especially when the average US citizen watches five hours of TV a day'. 'Why is it that pornography is considered to have this wild brain effect? Either it's because it's sexual, or because people are masturbating while watching it.' In his opinion, masturbation frequency is a big issue that has not been adequately addressed in current neuroimaging studies of porn use.

SEX, RELIGION, POLITICS AND MONEY: AREN'T THESE all the things we shouldn't discuss if we want to avoid trouble? And here I am talking about all of them! David told me trouble has indeed come his way, simply for being outspoken and challenging the validity of the concept of porn/sex addiction. He gets fairly regular hate mail and death threats from hardcore advocates of the concept. He is not the only one to have received such threats. This has also happened to sex researcher Dr Nicole Prause (see Chapter 5). She studied genital physiology for years, and it was only when she started doing research on porn use and reporting her findings that challenged some of the notions of the porn addiction concept that the threats began. 'It's a tough area. There are so many minefields,' she said to me. When I commented that she must have been brave to embark on this research topic, she laughed and said, 'Yes – and a little stupid!'

A common concern you'll hear is that if you watch a lot of porn, you'll eventually become unresponsive to your partner, and develop difficulties in responding sexually. But Nicole's work on female orgasms doesn't support this. She has found empirical evidence that porn use in women does not diminish sexual pleasure at all. In fact, it appears to do the reverse and actually enhance it. The women in her female orgasm study who reported the most porn use also reported experiencing the most intense orgasms. 'Orgasm is a reflex – the physiology is what it is. So there's no reason to think porn should alter it. This is really interesting.' She also found, in a recent study of couples, that those who watched more porn reported a stronger sexual urge just before their partnered sexual interactions in the lab, and that their porn use didn't impact at all on their sexual physiological responses; they showed no differences in their responses from those couples who didn't watch porn.

Nicole argues that people who watch a lot of porn have high sex drives. 'They engage with a lot of porn because they engage in a lot of sex. If they are distressed by this, there is typically a moral conflict,' she said. In her work she has found a lot of 'mixed affect' – that is, positive and negative emotions – associated with porn use. 'When people are asked about how they feel after watching porn alone, they report feeling sexually aroused, happy and excited, but also guilty, sad and anxious. But if they are watching it with a partner, they report the same positive emotions, in addition to feeling *less* sad and anxious.' She also points out that 'no one just watches porn – they masturbate to it'. This means that masturbation and orgasm are what are called the 'primary reinforcers', while porn is a 'secondary reinforcer'. 'There is some evidence that these two types of rewards/ reinforcers have overlapping but distinct neural representations,' Nicole said to me. This may be the reason why porn viewing habits alone have not been found to cause difficulties in sexual arousal or response with a partner. Masturbation habits, though, *may* generalise to sex with a partner, Nicole suggested. The conflation of masturbation habits with porn use will be discussed a little later.

FOR NOW, LET'S HEAD BACK TO THAT NEWLY PROPOSED ICD-11 diagnosis of 'compulsive sexual behaviour disorder' – what exactly is it, and what symptoms do you need to be given this diagnosis? As outlined by psychologist Shane Kraus and colleagues, it is characterised by a 'persistent pattern of failure to control intense, repetitive sexual impulse or urges, resulting in repetitive sexual behaviour over an extended period (e.g., six months or more) that causes marked distress or impairment in personal, family, social, educational, occupational or other important areas of functioning'. A diagnosis of compulsive sexual behaviour disorder can only be made if the repetitive sexual behaviours consume all aspects of a person's life, so that they neglect everything else – their work, health and personal care. They have to have made attempts to control it, and continue to engage in it despite consequences such as repeated relationship breakdowns. So, you can't be diagnosed with compulsive sexual behaviour disorder if you have a high sex drive that you can control and that causes you no distress or impairment in everyday functioning. The diagnosis also can't be given to people with high levels of sexual interest and behaviour (for example, masturbation) that are common in adolescents, even if the person suffers psychological distress. If the psychological distress is related to moral judgments or social disapproval about sexual impulses or behaviours, no diagnosis can be made. It's difficult to know the prevalence of this disorder given the inconsistent definitions and absence of community-based data, but it is estimated at somewhere around 3–6 per cent in adults. Interestingly, David argued that most people currently diagnosed as sex/porn addicts would not qualify (for this diagnosis). He also noted that the World Health

Organization has not yet formally approved the proposed diagnosis, and even if it does, it's unlikely to have any effect in the United States, the sex/porn addiction centre of the world, as they use the DSM and not the ICD for psychiatric diagnoses.

I wondered if people identified as having 'compulsive sexual behaviour disorder' have different brains. Would their brains display similarities to those of high porn consumers? This question has recently been addressed by neuroimaging research. Heterosexual males with 'problematic hypersexual behaviour' were recruited via meetings for 'sex addicts' and treatment facilities for 'sex addiction'. The men studied had each had an average number of 20 sexual partners in the previous six months, had sex on average three times a week, masturbated on average five times a week and viewed pornography five times per week. They were compared with a control group of males with the same level of education and income who did not report 'hypersexual' behaviours on screening tests. Men in the control group had an average number of two sexual partners in the previous six months, had sex on average less than once a week, masturbated on average once a week and viewed pornography twice a week.

Participants' brains were scanned while they viewed sexual images of naked women and sexual activities, and while they viewed 'non-sexual' images of water sport activities. Compared with the controls, the hypersexual men expressed greater sexual desire when viewing the sexual images and also showed greater activation in certain brain regions, including many that have been mentioned in this book already, such as the right frontal region (specifically the dorsolateral prefrontal cortex), and subcortical regions including the right dorsal anterior cingulate cortex, left caudate nucleus and right thalamus. These brain regions are known to be involved in motivation, reward processing and physiological responses – in other words, getting ready for sexual activity. The authors reported that the level of activation in frontal regions and the thalamus were directly correlated with the severity of the participants' sex 'addiction'. In other words, there was more brain activation in men who had higher scores on a hypersexual behaviour questionnaire. The authors of the study also reported that the functional brain changes observed were similar to those in people with substance addictions, so it is clear that these researchers are coming from the 'sex addiction is a true addiction' camp.

These same researchers did another study looking at structural brain differences in hypersexual men. They found that the temporal lobes of men who displayed 'problematic hypersexual behaviours' were reduced in volume (yes, once again, there is no doubting the temporal lobe's status as central to the sexual neural network), and also found reduced connectivity between these smaller temporal regions and other brain areas, including the left precuneus (involved in attention shifting) and the right caudate (involved in reward processing). The authors point out that people with addiction problems have trouble with shifting their attention and with reward-based behavioural learning;

these relate to the maintenance of addictive behaviours. I was surprised to see they referred to my own PhD studies in their discussion of the role of the temporal lobes in human sexual behaviour. They argue that the temporal regions are related to inhibition of the development of sexual arousal – in other words, they put the brakes on your sex drive. If there is damage or dysfunction in the temporal lobes (and as we know from earlier discussions, specifically the amygdalae within the temporal lobes), then these brakes become faulty and there is an alleviation of the inhibition which can potentially lead to hypersexuality (see Chapter 1).

Nevertheless, just like the neuroimaging research on frequent consumers of pornography, these studies are correlational. That means they only show a mutual relationship between two things. They don't tell us the cause. So we don't know if these brain differences predispose people to compulsive sexual disorder or are actually the result of it – or are actually related to something else altogether! Like high porn consumers, people with hypersexual disorders have been found to have high rates of depression and anxiety. They can also have features of attention deficit and hyperactivity disorder (ADHD) and obsessive-compulsive disorder. All these conditions are associated with changes to brain structure and function, so disentangling all of this is extremely complex.

David Ley also highlighted that none of the neuroimaging studies addressing this issue (of brain differences in so-called sex or porn addicts) have considered critical issues such as libido, frequency of masturbation and differences in sensation seeking – a personality trait involving the tendency to seek out new and intense sensations or experiences, even if they are risky, which in itself is associated with brain differences. There is still a lot more research to be done to disentangle all of these factors and determine how they might impact on brain structure and function. In a nutshell: just because brain differences are found in people with a high frequency of sex or porn use, it doesn't mean these differences are definitely the cause *or* the result of their sexual behaviours. There is much more research that needs to be done before it will be possible to answer the question 'can porn change your brain?'

ALTHOUGH THE RESEARCH IS NOT CONCLUSIVE, I wanted to talk to some people working on the front-line of providing treatments for people with excessive sexual behaviours to find out their opinions. Dr Vanessa Thompson is a sex therapist who has worked in private practice in Australia for a decade and sees a diverse range of people, including 'mainstream' clients, people with intellectual disability, and people who have committed sexual offences. She told me there is no 'typical' sex therapy client. She sees an equal number of males and females, ranging in age from school-aged children who have suffered sexual abuse to men in their eighties who need help with erectile problems.

I asked her about people with so-called 'sex addiction'. 'I've seen so many people who just love sex a lot – mainly men who definitely struggle to control their sexual urges,' she said, but she is not in the addiction camp. 'I don't work from an addiction perspective,' she clarified. 'When you treat an addiction you try to stop the behaviour. With sex addiction, you don't want them to leave therapy non-sexual. They don't want this, and neither do their part-ners.' She follows what she calls a 'sexual control' model, 'to teach them to understand their thoughts and behav-iours and get some control over them'. 'Sexual control', she explained, was different for everyone; for example, what is acceptable for a single guy in his twenties may not be for a married guy in his thirties. 'It's about them being in control of what they're doing. Not me telling them what's wrong with what they're doing or saying, "You can't do this or that."'

So how did she end up being a sex therapist? Vanessa said she had always been interested in sexual health and during her undergraduate degree, her first assignment was interviewing people on the university campus about whether they used condoms. She then worked in behav-iour intervention for people with intellectual disability and brain injury and described herself as 'the only person in the team who could say "penis" without laughing'. She ended up seeing all the clients who had problematic 'sexu-alised behaviours', such as exposing their genitals or public masturbation. She admitted that initially she had no idea what she was doing, so she tried to learn all she could by attending different courses around the country; she ended up doing a Master of Health Science (Sexual Health), qualifying as a sex therapist and then completing her PhD on 'sexual knowledge assessment tools for people with intellectual disability'.

Vanessa has provided therapy to people who have experienced a brain injury and resultant hyposexuality, hypersexuality or inability to orgasm. She couldn't recall any clients with paraphilias after brain injury but commented that it was difficult to assess whether people affected in this way had had 'unique interests' before their injuries; some may have already had a tendency to be outside the 'vanilla' – the conventional – realm in their sexual behaviours or preferences, she suggested. She recalled one of her first private practice clients who was referred for hypersexuality after brain injury, and who arrived at her office and immediately dropped his pants, revealing his penis to her shocked receptionist. He was a single man desperately seeking partners, but 'there was nowhere he could get his sexual needs met. Sex workers were too expensive for him.' Their therapy sessions involved making rules and a 'contract' that they agreed on, setting up a plan and structure that he could follow, and giving him advice on the best ways to find a sexual partner (certainly not by immediately dropping your pants!).

Vanessa's approach is very directive: she tells people what to try and gives them homework to do, then they report back to her what works or doesn't work for them. There is 'no touching or looking at bits' in her sessions, she emphasised. 'No Masters and Johnson here!' – in other words, her therapy doesn't include watching people have sex. At the core of her practice is an acknowledgment of individual differences. 'Everyone is different. There's no point me telling them they have to put their finger up their nose if they don't want to do it,' she said. Her favourite part of her work is 'seeing people succeed and reach their goals'. Her least favourite part is when clients 'divert away from talking about sex and start talking about who's doing the dishes … Relationship dynamics definitely come into play, but I'd rather they see a couples counsellor and just let me deal with the sex.'

I also spoke with Brian Russman, an American psychotherapist who has worked in the addiction field for nearly two decades. He tells me he is a 'Certified Sex Addiction Therapist' under the training scheme founded by Patrick Carnes. I was curious to learn what a 'certified sex addiction therapist' actually does. Brian is the deputy chief clinical officer at a specialised addiction treatment centre in Thailand; its website offers views of lush grounds with people meditating, being massaged and exercising by a pool. The music that played while the images rolled across my screen sounded hopeful, like it was specifically composed to make people think, '*This* will fix me.' It looked more like an expensive holiday resort than a treatment centre, and I wondered how the people who stayed there could afford it.

Brian estimated that in the six months that he had spent working in an outpatient clinic in Singapore, 50 per cent of his clients had presented for 'primary sex addiction'. He reflected that this may be an environmental issue, as the penalties for illicit drug use in Singapore are severe, so there are more sex, gambling and cyber addictions. At his workplace in Thailand, he estimated that 10–15 per cent of the people in residential treatment have a primary sex or cyber addiction.

'We're starting to see a huge tidal wave of people with sex and cyber addictions,' he said. 'Ten years ago, I rarely saw people with primary sex addiction, and never saw gaming disorder. These two addictions are now presenting much more for treatment, and it's the direct result of the internet. The vast majority of those with sex addiction have a component that is related to the internet, whether it's using porn or seeking partners through sites such as Tinder or Grindr.' He sees more males than females with sex addiction, at a ratio of about 4:1, and also mentioned that his patients often have paraphilias, typically exhibitionism or voyeurism, and more rarely fetishes for things like spectacles, feet or shoes.

People typically refer themselves to the Thai treatment centre where he works. 'Nine times out of ten, if they think they have a sex addiction, they do,' he said. He seemed genuinely excited about the recognition of 'sex addiction' in the ICD-11. He didn't refer to its official diagnostic label of 'compulsive sexual behaviour disorder', but perhaps for him and some others working in the field, these terms refer to the same behaviours, and so they use them interchangeably. He believes that formal recognition of the disorder would make life easier for those who have it and pave the way for more research. 'It's a catch 22 because if it is not a bona fide disorder then there's no money for research or treatment, which makes it hard to make the diagnosis official. There's also a lot of shame and stigma if it is not a legitimate diagnosis.'

I asked Brian how he treats his clients with 'sex addiction'. After his assessment, he makes an 'abstinence contract' with his client. 'The rule of thumb is 90 days,' he explained. 'No masturbation [or] sex with any partner; no caffeine, sugar, nicotine or alcohol. The hedonic set point is at such a high level. The rationale is to reset the brain.' This phrase, 'resetting the brain', came up several times in our conversation. It certainly is a powerful metaphor that is easy to comprehend, but I couldn't help wondering if a carefully designed and rigorous neuroimaging study would actually show any brain changes in response to abstinence from porn or sex. The jury is still out on the question of whether porn changes the brain, or if some people are just born with brains wired for porn. You don't really need to know the answer to this if you are just exploring if abstinence changes the brain. Brian acknowledged that for those who he sees as having 'sex addiction', complete abstinence for the rest of their lives is not the goal, although it would be for alcohol and drug addictions. He said that 'healthy sexual behaviours' are slowly reintroduced after the period of abstinence. 'Having said that, porn is not an option,' he clarified.

Coming as he does from the non-addiction camp, David Ley takes a different approach. He doesn't believe that porn 'changes' the brain. 'The brain is constantly changing,' he said to me. 'The crazy language about rebooting the brain after 90 days of abstinence? It's wild stuff.' In contrast to Brian's idea that a formal diagnosis of 'sex addiction' as a mental disorder would destigmatise the condition, David argues that the sex/porn addiction model and portrayal of these conditions as emerging from a brain issue – the 'brain-disease fallacy' of addiction, as it has been called – actually increases stigma and leads to people believing that these behaviours cannot be changed.

His approach to treating people who have previously been diagnosed with a sex or porn 'addiction' is also very different:

> If I walk into a doctor's office sneezing, they don't say to me, 'You've got a sneezing addiction.' They try to figure out what's underlying it – is it a virus, an allergy, a bacterial infection? That's what I do. I try to figure out what purpose or function the behaviour serves. Why is it a conflict?

He uses psychotherapy techniques including cognitive behavioural therapy, a thoroughly researched, empirically tested and effective psychological treatment for a variety of mental health conditions. He is firm in his view that 'as clinicians, we have an absolute ethical obligation to not promote unproven experimental treatments without acknowledging their severe limitations'.

David also had an interesting point about treatment and recovery from sex addiction. He told me about research that had shown that 95 per cent of people who self-identify as addicted to sex or porn actually get better on their own, without any treatment, within a year. Being older and having better adjustment in life are protective factors from developing any so-called behavioural (as opposed to substance use) addictions. This suggests that these 'behavioural addictions' are not a disease, he said, but 'reflect the need to adjust and accommodate to changes in life'.

THIS CHAPTER HAS BEEN A MERE INTRODUCTION TO the complexity of the highly controversial concept of porn and sex 'addictions'. Given that there is ongoing debate among experts and clinicians in the field about whether these are bona fide 'addictions', it is no surprise that our understanding of the neural basis of excessive porn use or compulsive sexual behaviour disorder is currently limited. Pre-existing brain abnormalities such as differences in the size of certain brain structures might make you more likely to watch a lot of porn, or predispose you to developing compulsive sexual behaviour disorder – or these brain differences could actually be the *result* of frequent porn use or excessive sex, or even due to something else entirely, like differences in sensation seeking or associated mental health conditions. I would have loved to do a neuropsychological assessment on my fellow train commuter, just to contribute a tiny piece to the puzzle – to understand the cognitive profile of someone who appeared to see nothing unusual about openly watching porn on his peak-hour train ride.

Who knows what the future holds for research into the neural correlates of frequent porn use and compulsive sexual behaviour? Will there ever be agreement among clinicians, scientists and other experts in the field about whether there is such a thing as sex or porn 'addiction'? What cultural factors are at play right now that impact on all of these issues? So many questions are raised, and it is only by conducting more methodologically rigorous scientific research that we will find answers. Hopefully some more researchers will be brave enough to take on the challenge.

10

BLAME IT ON MY BRAIN

Imagine you have a tumour growing in your frontal lobe. You become hypersexual, and begin to demand more sex from your long-term partner. Perhaps this reignites the passion you both felt in the first few months of your relationship and it brings you closer together. Alternatively, your mismatched libidos may frustrate and exhaust your partner, causing resentment and eventually leading to the relationship breaking down altogether.

In this scenario, whichever way it goes, your hypersexual behaviour is only directed towards your partner. It is contained to your bedroom, and its consequences are personal and specific to your relationship. Those consequences may be disastrous consequences, but they are restricted to you and your partner and do not directly impact others in your life. In contrast, consider the case of Gary (see Chapter 7). His right frontal lobe brain tumour resulted in hypersexuality that impacted dramatically on others. It led to him downloading child pornography and making sexual advances towards his stepdaughter – both of which are criminal offences. From his case report, we know he was ordered to complete a treatment program for 'sexual addiction', but there are no other comments about any legal proceedings.

We know that there have been rare cases of people who have committed sexual offences as a result of hypersexuality brought on by a neurological disorder or treatment. These people have faced the criminal justice system. But what happens when these people are charged with a sexual offence or assault? Do they blame their brains and, if so, how does the law respond? Are they deemed to be 'responsible' for their hypersexual behaviour? Are they punished?

The case of Isaac, a man with dementia who I met in my clinical practice, shows that there are no easy answers to these questions. Isaac was depressed. He was in his eighties, and he was lonely and tired. He missed his wife, who had died ten years before, and he missed working. He had loved being a high-school teacher, watching the kids in his class soak up the information he offered. He had been respected in his community and even served as president of a cultural organisation. Now he felt old and useless. He had so many medical issues, and took so many medications that didn't seem to help. He had been living with one of his sons, but when he started leaving pots burning on the stove and getting lost on his way home from the local shops, his family felt it was too hard to look after him. He moved into an aged care facility.

He had been living there for two years when his son contacted me, asking for an urgent neuropsychological assessment for his father. Isaac had been charged with an alleged sexual assault on his neighbour at the facility, a woman with severe dementia; he had been found by a staff member with his hands in her underpants. He had been taken to the police station and fingerprinted. His son had accompanied him and requested the police cease the interview when he realised his father's responses were confused and often incorrect. Isaac was charged with sexual assault, an apprehended violence order (AVO) was issued against him, and a request for a DNA sample was made.

Isaac had seen a geriatrician the week before his appointment with me, and the question of whether he had some type of dementia had been raised. In my assessment, he had a lot of trouble doing the 'frontal' executive tasks that I gave him, particularly those that required him to control his responses. For example, if I said a sentence with the last word missing and asked him to complete it with a word that *did not* fit, he just could not resist saying the word that *did* fit. He couldn't stop this automatic response. From the history of his cognitive decline that I got from his son, and the way he behaved with me – making comments such as, 'You are a naughty girl. You are teasing me. I am being silly' – it was clear that Isaac's frontal lobes were no longer doing their job of putting on the brakes in social situations. Documents from the aged care facility showed that staff had made increasingly frequent complaints about his sexually disinhibited behaviour; they had reported comments and demands that he had made, such as 'Give me a kiss'.

In my report, I advised that Isaac had probable vascular or behavioural-variant frontotemporal dementia. His sons were relieved, but also sad and worried. They gave my report to the legal aid solicitor who was helping them. Isaac effectively became a prisoner in his room due to the AVO: because his alleged victim was his neighbour, if Isaac left his room he was in breach of the AVO. He developed bed-sores and became even more depressed. He felt that other residents and staff had ostracised him. Eventually he was moved into the dementia unit at the same facility, and life became a little easier for him – he could safely leave his room with no legal ramifications. But Isaac's case shows how complex this issue is for everyone involved: the aged care staff, the police, the family of the victim, Isaac's family, and Isaac himself.

The aged care staff are trying to manage a challenging situation. They are bound by a duty of care (to both Isaac and other residents), and by laws that require them to report any incidents of 'unlawful sexual contact' or any sexual contact with a resident without consent. Would another staff member witnessing the incident have reported it, or would they have just shooed Isaac out of his neighbour's room and back to his own? How did the witness know if the neighbour had given consent or not? How is the idea of consent treated in residential homes where people have dementia? Staff did not know Isaac had dementia at the time of the alleged offence. If they had known, would they have handled the situation differently?

The police, faced with someone who has potentially committed a serious crime, are following the steps they are obliged to take in such incidents. But when the alleged perpetrator is an 80-year-old man with multiple medical conditions who is clearly confused, are there any alternatives to these standard practices? Do the police have any guidelines on how to manage elderly offenders with dementia? I'm not aware of any, but I know that in New South Wales, where I live, the government has allocated $10 million for a Justice Advocacy Service to help people with cognitive impairment navigate the criminal justice system, and to provide guidance to police, legal aid and the courts in responding to people with cognitive impairment such as dementia.

The family of Isaac's alleged victim, already living with the sad reality of her dementia, is now faced with the news that she was allegedly assaulted in a place where they thought she was safe. They are no doubt reeling, and must surely feel that someone should be brought to account or punished.

Isaac's sons are trying to reconcile their memories of their highly respected father with who he is now: a man with dementia, in the final stage of his life, who is now facing criminal charges. The myriad emotions they are dealing with must be exhausting.

And of course there is Isaac, a man with a brain that has no brakes due to a dementia that had started years before he walked into his neighbour's room, but wasn't diagnosed until after the alleged incident. Should he be considered responsible for his behaviour, and should he be punished? And, if so, how?

I met Isaac's son at the courthouse a year after we had

first met at Isaac's neuropsychological assessment. He introduced me to the solicitor who was running Isaac's case. 'I think it's a lay-down misère,' she said, meaning that she felt the outcome was certain. She explained that she would be arguing that Isaac was 'unfit to plead', and that she couldn't see the point of proceeding to court at all in cases such as Isaac's. 'He couldn't respond to any questions about the alleged incident,' she said. 'There is no purpose of punishment for a person with dementia. They really should be diverted out of the criminal justice system.' I nodded in agreement. Nevertheless, she admitted that she was unsure of the arguments the prosecution was going to present. She told us the process could take a few hours, as there were several other matters to be dealt with first, and that the actual hearing 'could take one minute or an hour'. I decided to stay and hear the outcome of Isaac's court case firsthand.

There's a lot of waiting that goes on in courts. The foyer was full of people nervously huddled in corners, talking softly with friends and family, staring blankly into their mobile phones or anxiously listening to their immaculately dressed lawyers. No doubt everyone was wondering when it would be their turn to face the magistrate who would decide their fate. I watched the magistrate as he deliberated over several other cases. He often spent time reading through documents while everyone in the court watched him in complete silence. I marvelled at his composure, the incredible level of responsibility he had, and his ability to process so much material on the spot and make life-changing decisions about people he had only just laid eyes on. He spoke eloquently and succinctly – not an 'um' to be heard. I wondered how much experience he'd had with people with dementia, and how this might influence his decision in Isaac's case.

There was no need for Isaac to be in court. His son was relieved that they had shielded him from all the worry and uncertainty over the previous year. 'He has no idea about any of this,' he said to me. 'I'm glad he doesn't have to sit here. He was a pillar of society, a well-respected teacher. Surely this can't be the legacy of his life?' He proudly showed me some recent video footage of Isaac on his phone. He looked relaxed as he stared directly into the camera and spoke in his native language. Isaac's son agreed that Isaac was much happier in the dementia unit, noting that the staff were 'much more kind, flexible and forgiving'. Isaac was still disinhibited, but there had been no further complaints about his behaviour since he moved to the dementia unit.

I asked Isaac's son if his father's disinhibition was still sexual in nature, and he nodded and gave me two examples of incidents that had occurred during his recent visits. On one occasion a movie was playing in the communal lounge and there was a scene of a couple kissing. Isaac leaped out of his seat and announced to the room, 'They are kissing! Look, they are kissing!', raising his hands above his head and clapping. His son was shocked at how excited he became. Another time Isaac pointed to his two slippers that were on the floor of his cupboard, one placed on top of the other. He remarked, 'Look, they are humping, like rabbits!'

The son laughed and shook his head. 'For him to interpret two slippers in that way? Wow. I just couldn't believe it.' Isaac really *did* have sex on the brain.

In the end, a year of worry and four hours of waiting in court were resolved in a one-minute judgment. The magistrate said the evidence from the reports about Isaac was 'overwhelming with respect to his lack of mental capacity' and that 'the only conclusion to reach is that he is not fit to be tried'. He dismissed the case of sexual assault, and then the prosecution raised the issue of the AVO. Isaac's lawyer spoke for ten seconds. 'If it's to anyone's comfort he is now in a dementia unit, Your Honour, essentially under lock and key' – nowhere near his former neighbour in the facility. The magistrate paused, then replied, 'I suppose it's the same conclusion. There is no prospect of him understanding or complying with such an order. Dismissed.'

I nodded to Isaac's son across the aisle as we stood up and headed for the door. The solicitor bowed to the magistrate and darted out behind us.

'Is that it, then?' Isaac's son asked.

'Yes,' she answered. 'As I said – a lay-down misère.'

As we walked to the court's car park, Isaac's son seemed tired but very relieved. 'So much worry, and what for?' he asked. 'It should never have got to this.' We shook hands, and on my way home I wondered how many other cases like Isaac's were going through the criminal courts right now, causing angst for all the families involved.

SOME OF THE PATIENTS I HAVE ALREADY DISCUSSED IN Chapter 7 were charged, as Isaac was, but were then imprisoned for their criminal behaviour despite evidence of their neurological condition being presented to the court. Debbie, who had multiple sclerosis (MS) and developed a range of paraphilias, ended up dying in her jail cell. Her case report by Neil Ortego and colleagues stated that:

> A possible relationship between the patient's MS and her bizarre behavior was ignored. She was thought to be cognitively intact with a 'normal IQ,' and thus in full awareness and control of her actions. Based on this line of reasoning, she was given the maximum sentence and died in the jail cell. Her neurological tragedy was unnecessarily exaggerated by the disregard for the relationship between her MS and her altered behavior.

Todd, who had part of his right temporal lobe removed to treat his temporal lobe epilepsy (see Chapter 7), also ended up in prison for his child pornography offence. The authors of his case study argued strongly that he should not have been charged or imprisoned:

> Among more than 35 cases of pedophilia associated with neurological disorders, an arrest was only reported in two cases … In contrast, our patient was prosecuted by federal authorities and now serves a 19-month prison sentence for downloading

pornographic images of children. The KBS [Klüver-Bucy syndrome] was a critical factor in driving his hypersexuality. In light of this mitigating factor, was he criminally responsible? Did his behavioral actions warrant imprisonment? We believe the answer is no to both questions.

We learn more about Todd's court case and sentencing from Oliver Sacks' posthumously published book. Sacks states that the judge agreed that he 'could not be held accountable for having Klüver-Bucy syndrome', but argued that he was culpable 'for not speaking sooner about the problem to his doctors, who could have helped, and for persisting for many years in behavior that was injurious to others'. Todd's crime, the judge said, 'was not a victimless one'.

In the case of Gary, also discussed in Chapter 7, there was a clear association between his tumour and hypersexual behaviour: the behaviour stopped when the tumour was removed, and started again when the tumour grew back. This is evidence of causality. We know from abundant neuropsychological and neuroimaging research that the orbitofrontal region, where Gary's tumour was located, is crucial in regulating social behaviour. If this brain region is damaged during childhood, it can cause difficulties in learning and understanding social norms and rules. Symptoms of orbitofrontal damage include poor social judgment, lack of empathy and remorse, reduced impulse control, and difficulty interpreting people's emotions and mental states. If you injure this brain region as an adult, you can manifest what has been termed 'acquired socio-pathy'. Famous patients such as 'EVR' and Phineas Gage have highlighted how this type of brain damage can have a devastating impact on a person's ability to function in the social world, resulting in unemployment and family breakdown. In some people with orbitofrontal brain damage there can be a dissociation between 'knowing' and 'doing', in that people can know what they are doing is wrong, but are unable to stop themselves acting in certain ways. In other words, their 'moral knowledge' is intact, but they have a loss of impulse control, or a failure of their brakes. The authors of Gary's case study, neurologists Jeffrey Burns and Russell Swerdlow, stated that:

> The orbitofrontal disruption likely exacerbated a preexisting interest in pornography, manifesting as sexual deviancy and pedophilia. To our knowledge, this is the first description of pedophilia as a specific manifestation of orbitofrontal syndrome … Our patient could not refrain from acting on his pedophilia despite the awareness that this behavior was inappropriate.

In other words, Gary knew what he was doing was wrong, but he could not stop himself.

The authors of Debbie's, Todd's and Gary's published case studies all argued that their patients should not have been held responsible for their sexual offences. They felt that the law disregarded the causal link between brain conditions and criminal sexual behaviour and that this crucial factor should have been considered. No one can be considered 'responsible' for developing multiple sclerosis or temporal lobe epilepsy, or for growing a brain tumour. These people have already suffered a 'neurological tragedy', and the symptoms of hypersexuality only compound this. Isn't that punishment enough?

But there were victims, too, in these cases, and in all cases those victims were children. Isn't there a risk that people like Debbie, Gary and Todd will commit more offences? Surely their crimes are too significant for us to let these people off with a warning? And where do we draw the line? Of course 'your brain made you do it' – our brains make us do everything! If charges are dropped because paedophilia occurred in the context of a brain tumour, couldn't *all* paedophiles just claim that their brains are wired differently and therefore they are not responsible for their crimes? Doesn't this raise the question of whether 'free will' actually exists? Is there a risk that anyone committing a sexual offence could just blame their brain? As you can see, these cases of hypersexuality arising from brain disorders raise some very complex issues.

In the case of the criminal justice system, the two obvious and interrelated questions are: (1) are they responsible for their criminal behaviour; and (2) should they be punished – and if so, how? Legal, ethical and philosophical scholars have spent entire careers researching such issues, so it is far beyond the scope of this chapter to adequately address them, but I will highlight some relevant cases that show how legal systems in Australia and other countries are dealing with this. Specifically, I will describe what happened to two men with Parkinson's disease who were charged with sexual offences that they committed while taking prescribed medications for their condition – medications that were directly responsible for their hypersexuality.

IF YOU DEVELOPED PARKINSON'S DISEASE, YOU WOULD no doubt be desperate for some relief from your motor symptoms. You would gladly take any medications that your doctor prescribed if they stopped you from shaking, shuffling and freezing while you walked. Just as Steven (see Chapter 2) and Ray (see Chapter 4) were, you would be thrilled if those drugs were effective and you could walk freely again. Nevertheless, the joy you felt might be tainted if you developed a known side effect of the medication: hypersexuality.

If your prescribed medication caused hypersexuality that led to criminal behaviour, should you be punished? Not according to the judges of two such cases, from the United Kingdom and Australia. 'Parkinson's caused teacher's child porn habit, judge rules' was the headline of a 2008 story in one UK newspaper. Philip Carmichael, a 58-year-old retired school principal, had been prescribed dopamine replacement therapy to relieve symptoms of his Parkinson's. During the period in which he took the medication, he downloaded 8000 child pornography images. Police also found one image on his computer that Carmichael had downloaded before starting this medication; Carmichael claimed he did not know where it came from. The existence of this single 'pre-medication' image raises the question of whether the medication unmasked a latent or pre-existing tendency. The judge in Carmichael's case disregarded this notion as this image was entirely separate from the bulk of the downloaded images. Judge Mary-Jane Mowat stated that:

> This is a very distressing case. To say that he was to blame would be a complete denial of the reality of the evidence that I see. He was not only an ill man at the time, but a man whose medication can be described as ultimately responsible for the committal of these offences.

Judge Mowat noted that Carmichael was now under medical supervision and had a loyal social support network; she considered it unlikely that he would reoffend. It was not deemed necessary to ban him from working with children.

There was a similar case in Australia of a Tasmanian politician who developed hypersexuality after taking prescribed medications for his Parkinson's disease. Over two years, Terry Martin became increasingly obsessed with sex. He spent approximately $150 000 on 162 different sex workers on 506 occasions, and recorded details of his encounters on a spreadsheet. His sexual preferences expanded from his pre-medication heterosexuality to engaging with transsexual and male prostitutes. Eventually, his hypersexuality led to criminal behaviours. He was found guilty of several criminal offences for having sexual intercourse with a person under the age of 17 years, and for producing and possessing child pornography. The judge in the Tasmanian Supreme Court case found a 'direct causal link between the medication prescribed for Mr Martin's Parkinson's disease and the offending ... he would not have committed any crimes if he had not taken those drugs'. This was supported by two observations made by medical expert witnesses in the trial: first, that Martin was unable to control his behaviour, even when he became aware of the link between his behaviour and his medication; and second, his hypersexual behaviours ceased when he stopped taking the medication. Martin received prison sentences but they were suspended.

A critical question raised by these two cases is: did these men know that their medications could induce hypersexuality and, if so, how long did they know of this link before their illegal behaviours began? Did they continue to take their medications even while they knew about the side effects? Maybe they had been warned by their doctors and knew of the possible link, but had no reason to suspect that it would cause them to engage in criminal behaviour. Should they have foreseen those consequences? There is plenty of evidence that dopamine replacement treatments can cause hypersexuality, but it is not clear if they specifically cause paedophilic interest. The question of whether these medications can unmask a previously unexpressed sexual desire has also been raised, and a case reported in *The Journal of Sexual Medicine* suggests that this is possible.

Norman was 67 years old and had been diagnosed with Parkinson's disease seven years earlier. His doctors had tried a few different medications to ease his motor symptoms and juggled the doses over the years. After increasing the dose of one of his medications (Pramipexole), his wife rang his doctor's office in tears. She described a profound change in Norman's sexual behaviour. He had been shy and reserved, and they had previously had sex approximately once a week, but since his medication increase he wanted sex daily. It wasn't just the ramping up of the frequency, though. His wife reported a change in how he wanted it and what he said during it: he had developed an extreme preference for anal sex, and would vocalise 'unusual obscenities' during it. He had never expressed this desire or requested this type of sex in over 40 years of marriage.

The doctor scheduled an appointment. Norman initially denied any changes, but when his wife confronted him, he said that he assumed his requests were 'unusual compared with his previous sexual experiences with her', but according to the case report, he said that 'these were practices that he secretly desired when he was younger', and that 'now he felt somehow less ashamed to put his desire into practice'.

The Pramipexole was ceased, and after 30 days his sexual behaviour returned to its usual pattern. When asked about his change in sexual behaviour at a subsequent review, he showed insight into how his drug-induced hypersexuality now seemed to him 'inadequate and unacceptable' in the context of his marriage.

Was Norman's sudden interest in anal sex an authentic expression of the 'real' Norman that the medication had 'unmasked' and finally allowed him to express after decades of repression, or was this a *de novo* medication-induced desire that was not really him at all? In the case of Philip Carmichael, did the single child pornography image that he downloaded before he started his medication constitute a 'latent tendency'? It is impossible to know. Carmichael denied any knowledge of the pre-medication image. He'd had an unblemished career as a school principal, and there were no previous reports of any paedophilic interests or behaviour, but only Carmichael knows his true self. The judge in his case certainly did not believe his criminal behaviour reflected a previously hidden paedophilic interest. Carmichael's own statement that the medication 'turned him into a paedophile' suggests that his criminal behaviour did not reflect an unmasked latent tendency, but was something entirely foreign that became impossible to stop.

WHEN JUDGES DETERMINE WHAT AN APPROPRIATE sentence or punishment will be for a particular offence and offender, they have to consider myriad factors. Some of these factors are 'general deterrence' (discouraging criminal behaviour through fear of punishment), 'future dangerousness' (considering how likely it is that the person will offend again), and 'custodial hardship' (considering whether the offender will suffer more than usual in prison).

Another factor that gets taken into account in sentencing is whether or not the person has at the time of sentencing, or had at the time of the offence, a 'mental impairment', meaning any type of restricted mental or intellectual functioning, whether it is temporary or permanent, or any mental disorder or abnormality. The impairment doesn't need to be a diagnosed mental illness, nor does it have to be severe to be taken into account by a judge. In Australia, a set of principles known as the 'Verdins principles' usually guides a court's sentencing in these cases; the principles arose from a decision made by the Supreme Court of Victoria Court of Appeal in 2007. They outline five ways that mental impairment can affect a sentence:

1 It can reduce a person's 'moral culpability' (or their responsibility from a moral perspective) in a number of ways. For example, impaired mental functioning can mean a person cannot think clearly, exercise appropriate judgment or appreciate that what they are doing is wrong.

2 It can influence the type of sentence that can be imposed and the conditions under which it can be served.

3 It can mean that deterrence may not be relevant in sentencing. For example, an offender who is not fully able to learn from a court's judgment is unlikely to

be deterred by a prison sentence from committing another crime.

4   It can increase the hardship suffered by a person in prison.

5   It may justify a less severe sentence if there is a risk that prison could adversely affect the person's mental health.

In the cases of Philip Carmichael and Terry Martin, the two offenders with Parkinson's disease, the judges felt that their moral culpability was reduced. Therefore their punishment – or, rather, the lack of formal punishment – reflected this. The judges also made references to other factors. In the case of Carmichael, Judge Mowat clearly considered that 'future dangerousness' was not an issue and explicitly stated that Carmichael was unlikely to reoffend, so there was no need to engage a probation service. In Martin's case, Justice David Porter stated that the case was 'not an appropriate vehicle for giving expression to the factor of general deterrence' and that 'individualised approaches are called for in these circumstances'. Custodial hardship was also addressed in the sentencing remarks, considering Martin's social isolation, previous suicide attempt and plans to move interstate to access neurosurgical treatment (deep brain stimulation). Justice Porter stated that 'imprisonment is ordinarily the only appropriate penalty ... However this is by no means an ordinary child pornography case.'

CASES OF HYPERSEXUALITY DUE TO NEUROLOGICAL conditions have not only featured in criminal trials within the justice system. Compensation cases have also been reported. In France, Didier Jambart was awarded damages and received €197 000 (AU$240 300) in compensation from GlaxoSmithKline for taking the Parkinson's disease drug ropinirole, sold as Requip, that made his life 'hell'. Within two years of beginning to take Requip, Jambart had transformed from a well-respected member of his community – he was a former bank manager and local councillor – into a man seemingly addicted to sex and gambling. He sold his children's toys, and stole from neighbours, friends and colleagues to fuel his gambling addiction; sexually, he engaged in a frantic search for gay sex and advertised himself on the internet to arrange encounters, one of which resulted in him allegedly being raped.

Jambart says the first he knew of the links between Requip and compulsive behaviours was when he found a website that warned of side effects in 2005. He ceased taking his medication and his behaviour returned to normal, yet warnings about Requip's side effects were only made public in 2006. The court found that there was 'serious, precise and corroborated' evidence to blame Jambart's transformation on Requip.

Didier Jambart was not the first patient with Parkinson's disease to sue a drug maker over their impulse control disorder. In a Minneapolis court in 2008, a man who developed a gambling addiction after taking pramipexole (in this case sold under the name Mirapex) was awarded

US$8.2 million for gambling losses and punitive damages against the pharmaceutical companies Pfizer and Boehringer Ingelheim. As discussed in Chapter 2, in Australia Pfizer agreed to settle a compensation claim by 160 patients who took the dopamine treatment drug Cabaser and developed gambling addictions or hypersexuality. A Salvation Army financial counsellor reported, 'There are people I'm sure who have committed suicide over this,' she said. 'They haven't known it was the drug doing it. There are families that have been ripped apart.'

People with Parkinson's disease are not the only patients who have received payouts for hypersexuality in the context of a brain condition. Alissa Afonina was a high-achieving high-school student in Canada. After she sustained a traumatic brain injury in a car accident, she became 'lethargic, disruptive, unable to follow course content and socially isolated' and dropped out of school. It was reported that she showed 'no impulse control, made inappropriate sexual comments, and could not follow through with tasks'. She started working as a stripper, and then as a dominatrix in sadomasochistic role-playing sessions with male clients. Her lawyers argued that her decision to do this type of work showed a 'lack of correct thinking', and that she had taken an unnecessary risk due to a loss of cognitive function from her brain injury. The judge noted that she had not acted to minimise her risks in her work, such as by implementing an alarm system or safety measures; this was cited as evidence of diminished judgment and frontal lobe damage. The judge concluded that Afonina's brain injury led to her being unable to cope normally and impaired her potential earning power, and that if she had not suffered a brain injury she would have been capable of completing a tertiary qualification. She was awarded damages that took into account future capacity loss, cost of future care, pain and suffering.

IT IS IMPORTANT TO BE AWARE THAT NOT EVERYONE with a neurological disorder experiences a change in their sex drive or behaviour, and even if they do, it is rare that it is dramatic enough to lead to criminal behaviours and legal proceedings like those described in this chapter. As I've noted throughout this book, *hypo*sexuality – a lack of or diminished sex drive – and reduced frequency of sexual behaviour is actually far more common than *hyper*sexuality as a sexual outcome for people with neurological disorders. But because hyposexuality is not associated with any criminal behaviour, all the cases in this chapter involve hypersexuality. I've highlighted these rare cases because they demonstrate the links between sex, our brains and the law, giving us insights into how legal systems are currently dealing with these complex issues.

That said, I believe that cases like Isaac's will become more common given our ageing population and the increasing incidence of dementia. People with dementia living in aged care facilities still have sex drives and sex lives, and this needs to be acknowledged. How they express their sexual lives, and who they express them with, can become very complicated issues in such an environment. Sex is one of the fundamental human behaviours, but it *is* complicated. That's what makes it so interesting.

AFTERWORD

I had not long finished writing this book when I heard something shocking from the daughter of a patient. I'd just spent about 20 minutes talking with Edith, a woman in her mid-sixties. Edith walked with a frame; she was slow to respond to my questions and often answered by going completely off topic. For example, when I asked her how her memory was going, she replied, 'I fell twice on my way here' with a big smile on her face.

Edith adamantly denied any cognitive changes or health concerns, but when I spoke with Edith's adult daughter alone, she described the severe cognitive and physical decline that Edith was experiencing. She had trouble walking and swallowing, and was incontinent. She didn't seem to understand what people were saying to her, and her text messages made no sense at all. It was when she began to talk about how Edith had become more sexually active over the last two years that the most alarming information came to light. On Edith's phone, her daughter had found sexually explicit text messages and photos that Edith had sent to men online. One text message had frightened her: it detailed a plan for a man to come to their house and rape her. Not Edith, but her *daughter*.

'Sorry?' I gasped.

Edith's daughter confirmed what I thought I'd heard: Edith had arranged for a man to rape her daughter. I held my breath. Edith's daughter wiped away tears as she told me that she had gone to the police, but they couldn't offer much help as her mother had consented to the man's visit. She threatened the man with police action, and thankfully nothing eventuated. Not surprisingly, the relationship between Edith and her daughter had become very strained; I noticed that they didn't look at or speak to each other at all. Her daughter confided that she couldn't manage the demands of caring for Edith anymore and was arranging for her to go to an aged care facility.

I spent the next few hours with Edith, doing various tests that showed she had severe cognitive difficulties. She often just stared at me and smiled. It was hard to imagine that this smiling, cognitively impaired woman could plan such a horrific assault on her daughter.

There was no doubt that Edith had behavioural-variant frontotemporal dementia, possibly the type that can be associated with motor neurone disease. This news was no surprise to her daughter, but it didn't erase the pain of the tragedy that had almost unfolded. I doubted Edith's daughter would ever forgive her mother, even knowing that Edith's brain was ravaged by a neurodegenerative disease.

What I have written about in this book is not just juicy trivia or fascinating facts about brain bits that play havoc with our sex lives. There can be devastating consequences for the family members of people who undergo such changes. I already knew this from some of the other cases I have described in this book, but something about Edith and her daughter really cemented this. They stuck in my mind long after I had sent off my report and packed Edith's file away in my filing cabinet.

Throughout this book, the case studies show that there are many parts of the brain that make up the sexual neural network, and that a disruption to any of those parts through brain injury or disease can have dramatic consequences. But the case studies and related research also raise so many questions. Why do only some people with injuries or disease affecting parts of the sexual neural network show sexual changes? Are people who develop compulsive behaviours in relation to porn or sex born with different brains, or do these behaviours change their brains? Why do only some people with paedophilia go on to commit sexual offences? I can only hope that raising these questions will stimulate further research that will eventually answer them.

Bringing these case studies out into the open is the first step in acknowledging that these things actually occur, even if they only occur rarely. By sharing these I hope I have made a small contribution to our understanding of the incredible complexity of sex, love and the brain. I am not a sex therapist and cannot advise on how to treat the sexual changes I describe, but if you are reading this and are experiencing troubling sexual changes associated with a neurological condition, I urge you to discuss them with your treating neurologist or GP, and seek out a local sex therapist.

As I wrote this book, I realised that although I love my work as a clinical neuropsychologist, I'm hungry for something more – something beyond assessing and reporting on people's cognitive functions. I've considered studying psychotherapy, sex therapy and even law. It is the last option that I keep coming back to; I'm fascinated by the complex questions of criminal responsibility that are raised by some of the cases, including Edith's.

Whatever path I take, there is no doubt that this book will have played a role in illuminating it. I hope you enjoyed reading it as much as I enjoyed writing it.

FURTHER READING

This list includes all the published case studies discussed in each chapter, in addition to a mix of relevant published scientific research studies, online science and newspaper articles, books, and other relevant guidelines and websites. It is not exhaustive, but is intended to provide the interested reader with further material relevant to each chapter.

1 YOUR SEXIEST BRAIN BITS

Baird, A.D., Wilson, S.J., Bladin, P.B., Saling, M.M., & Reutens, D.C. (2004). The amygdala and sexual drive: Insights from temporal lobe epilepsy surgery. *Annals of Neurology*, *55*(1), 87–96.

—— (2007). Neurological control of human sexual behaviour: Insights from lesion studies. *Journal of Neurology, Neurosurgery, and Psychiatry*, *78*(10), 1042–1049.

Brown, S., & Schäfer, E.A. (1888). An investigation into the functions of the occipital and temporal lobes of the monkey's brain. *Philosophical Transactions of the Royal Society of London. B*, *179*, 303–327.

Diamond, J. (1997/2015). *Why is Sex Fun? The evolution of human sexuality*. London: Weidenfeld & Nicolson.

Klüver, H., & Bucy, P.C. (1939). Preliminary analysis of functions of the temporal lobes in monkeys. *Archives of Neurology & Psychiatry*, *42*, 979–1000.

Lilly, R., Cummings, J.L., Benson, F., & Frankel, M. (1983). The human Klüver-Bucy syndrome. *Neurology*, *33*(9), 1141–1145.

Marlowe, W.B., Mancall, E.L., & Thomas, J.J. (1975). Complete Klüver-Bucy syndrome in man. *Cortex*, *11*(1), 53–59.

Ogden J. (2012). HM, the man with no memory. *Psychology Today*, 16 January 2012, <www.psychologytoday.com/au/blog/trouble-in-mind/201201/hm-the-man-no-memory>.

Terzian, H., & Dalle Ore, G. (1955). Syndrome of Klüver and Bucy, reproduced in man by bilateral removal of the temporal lobes. *Neurology*, *5*(6), 374–380.

2 'GIVE IT TO ME BABY' OR 'NOT TONIGHT, DARLING'

Ahmed, R., Kaizik, C., Irish, M., Mioshi, E., Dermody, N., Kiernan, M.C., … Hodges, J.R. (2015). Characterizing sexual behavior in frontotemporal dementia. *Journal of Alzheimer's Disease*, *46*(3), 677–686.

Aull-Watschinger, S., Pataraia, E., & Baumgartner, C. (2008). Sexual auras: Predominance of epileptic activity within the mesial temporal lobe. *Epilepsy & Behavior*, *12*(1), 124–127.

Blumer, D. (1970). Changes of sexual behavior related to temporal lobe disorders in man. *Journal of Sex Research*, *6*(3), 173–180.

Janszky, J., Ebner, A., Szupera, Z., Schulz, R., Hollo, A., Szücs, A., & Clemens, B. (2004). Orgasmic aura – a report of seven cases. *Seizure*, *13*(6), 441–444.

Ozkara, C., Ozdemir, S., Yılmaz, A., Uzan, M., Yeni, N., & Ozmen, M. (2006). Orgasm induced seizures: A study of six patients. *Epilepsia*, *47*(12), 2193–2197.

Rémillard, G.M., Andermann, F., Testa, G.F., Gloor, P., Aube, M., Martin, J.B., … Simpson, C. (1983). Sexual ictal manifestations predominate in women with temporal lobe epilepsy: A finding suggesting sexual dimorphism in the human brain. *Neurology*, *33*(3), 323–330.

Spencer, S.S., Spencer, D.D., Williamson, P.D., & Mattson, R.H. (1983). Sexual automatisms in complex partial seizures. *Neurology*, *33*(5), 527–533.

Stoléru, S., Fonteille, V., Cornélis, C., Joyal, C., & Moulier, V. (2012). Functional neuroimaging studies of sexual arousal and orgasm in healthy men and women: A review and meta-analysis. *Neuroscience & Biobehavioral Reviews*, *36*(6), 1481–1509.

Wise, N., Frangos, E., & Komisaruk, B.R. (2016). Activation of sensory cortex by imagined genital stimulation: An fMRI analysis. *Socioaffective Neuroscience & Psychology*, *6*, 31481.

―――― (2017). Brain activity unique to orgasm in women: An fMRI analysis. *Journal of Sexual Medicine*, *14*(11), 1380–1391.

Zhang, S., Dissanayaka, N., Dawson, A., O'Sullivan, J.D., Mosley, P., Hall, W., & Carter, A. (2016). Management of impulse control disorders in Parkinson's disease. *International Psychogeriatrics*, *28*(10), 1597–1614.

3 SEXUAL SIDE EFFECTS OF SEIZURE SURGERY

Asadi-Pooya, A.A, & Rostami, C. (2017). History of surgery for temporal lobe epilepsy. *Epilepsy & Behavior, 70*(A), 57–60.

Baird, A.D., Wilson, S.J., Bladin, P.B., Saling, M.M., & Reutens, D.C. (2002). Hypersexuality after temporal lobe resection. *Epilepsy & Behavior, 3*(2), 173–181.

—— (2003). Sexual outcome after epilepsy surgery. *Epilepsy & Behavior, 4*(3), 268–278.

Bladin, P.F., Wilson, S.J., Saling, M.M., Kincade, P., McIntosh, A., & O'Shea, M.F. (1999). Outcome assessment in seizure surgery: The role of postoperative adjustment. *Journal of Clinical Neuroscience, 6*(4), 313–318.

Rippon, G. (2019). *The Gendered Brain: The new neuroscience that shatters the myth of the female brain*. London: Bodley Head.

Ritchie, S.J., Cox, S.R., Shen, X., Lombardo, M.V., Reus, L.M., Alloza, C., ... Deary, I.J. (2018). Sex differences in the adult human brain: Evidence from 5216 UK Biobank participants. *Cerebral Cortex, 28*(8), 2959–2975.

Wheeling, K. (2015). The brains of men and women aren't really that different, study finds. *Science*, 30 November, <www.sciencemag.org/ news/2015/11/brains-men-and-women-aren-t-really-different-study-finds>.

Wilson, S.J., Bladin, P.F., & Saling, M.M. (2001). The 'burden of normality': Concepts of adjustment after surgery for seizures. *Journal of Neurology, Neurosurgery, and Psychiatry, 70*, 649–656.

4 STIMULATION, SHUNTS AND PSYCHOSURGERY

Conley, G. (2016). *Boy Erased: A memoir*. New York: Riverhead Books.

de Boisanger, L., & Kaliaperumal, C. (2018). Letter: Sex after neurosurgery – is it safe? *Neurosurgery, 82*(2), E67–E68.

Demetriades, P., Rickards, H., & Cavanna, A.E. (2011). Impulse control disorders following deep brain stimulation of the subthalamic nucleus in Parkinson's disease: Clinical aspects. *Parkinson's Disease, 2011*, 1–9.

Gorman, D.G., & Cummings, J.L. (1992). Hypersexuality following septal injury. *Archives of Neurology, 49*(3), 308–310.

Heath, R.G. (1964). Pleasure response of human subjects to direct stimulation of the brain: Physiologic and psychodynamic considerations. In R.G. Heath (ed.), *The role of pleasure in behavior* (pp. 219–243). New York: Harper & Row.

—— (1972). Pleasure and brain activity in man: Deep and surface electroencephalograms during orgasm. *Journal of Nervous and Mental Disease, 154*(1), 3–18.

Kesey, K. (1962). *One Flew Over the Cuckoo's Nest*. New York: Viking.

Moan, C.E., & Heath, R.G. (1972). Septal stimulation for the initiation of heterosexual behavior in a homosexual male. *Journal of Behavior Therapy and Experimental Psychiatry*, *3*(1), 23–26.

Rieber, I., & Sigusch, V. (1979). Psychosurgery on sex offenders and sexual 'deviants' in West Germany. *Archives of Sexual Behavior*, *8*(6), 523–527.

Schmidt, G., & Schorsch, E. (1981). Psychosurgery of sexually deviant patients: Review and analysis of new empirical findings. *Archives of Sexual Behavior*, *10*(3), 301–323.

5   CAN SEX CHANGE YOUR BRAIN?

Blaxendale, S. (2004). Memories aren't made of this: Amnesia at the movies. *British Medical Journal*, *329*, 1480–1482.

Brain Aneurysm Foundation. (2019). Statistics and facts. Brain Aneurysm Foundation, < https://bafound.org/about-brain-aneurysms/brain-aneurysm-basics/brain-aneurysm-statistics-and-facts/ >.

el Gaddal, Y.Y. (1989). De Clérambault's syndrome (erotomania) in organic delusional syndrome. *British Journal of Psychiatry*, *154*(5), 714–716.

Foreman, P.M., Griessenauer, C.J., Selim, M.H., Searls, D E.C., Safdar, A., Kasper, E.M., ... Thomas, A.J. (2016). Sexual activity as a trigger for intracranial hemorrhage. *Acta Neurochirurgica*, *158*(1), 189–195.

Inzitari, D., Pantoni, L., Lamassa, M., Pallanti, S., Pracucci, G., & Marini, P. (1997). Emotional arousal and phobia in transient global amnesia. *Archives of Neurology*, *54*(7), 866–873.

Kaplan, H.S. (1979). *Disorders of sexual desire and other new concepts and techniques in sex therapy*. New York: Brunner/Mazel.

Maloy, K., & Davis, J.E. (2011). 'Forgettable' sex: A case of transient global amnesia presenting to the emergency department. *Journal of Emergency Medicine*, *41*(3), 257–260.

Masters, W.H., & Johnson, V.E. (1966). *Human Sexual Response*. Boston: Little, Brown.

Reynolds, M.R., Willie, J.T., Zipfel, G.J., & Dacey Jr., R.G. (2011). Sexual intercourse and cerebral aneurysmal rupture: Potential mechanisms and precipitants: A review. *Journal of Neurosurgery*, *114*(4), 969–977.

Shorvon, S.D. (2011). The causes of epilepsy: Changing concepts of etiology of epilepsy over the past 150 years. *Epilepsia*, *52*(6), 1033–1044.

Sieveking, E. (1858). *On epilepsy and epileptiform seizures: Their causes, pathology and treatment*. London: John Churchill.

Simons, J.S., & Hodges, J.R. (2000). Transient global amnesia. *Neurocase*, *6*(3), 211–230.

Stoléru, S., Fonteille, V., Cornélis, C., Joyal, C., & Moulier, V. (2012). Functional neuroimaging studies of sexual arousal and orgasm in healthy men and women: A review and meta-analysis. *Neuroscience & Biobehavioral Reviews, 36*(6), 1481–1509.

Tissot, S.A.D. (1767). *Onanism: Or, a treatise upon the disorders produced by masturbation* (A. Hume, Transl.). London: A. Hume.

6 WHAT'S LOVE GOT TO DO WITH IT?

Baird, A., & Thompson, W.F. (2018). When music compensates language: A case study of severe aphasia in dementia and the use of music by a spousal caregiver. *Aphasiology, 33*(4), 449–465.

Eisenberger, N.I. (2012). The neural bases of social pain: Evidence for shared representations with physical pain. *Psychosomatic Medicine, 74*(2), 126–135.

el Gaddal, Y.Y. (1989). De Clérambault's syndrome (erotomania) in organic delusional syndrome. *British Journal of Psychiatry, 154*(5), 714–716.

Fisher, H.E., Brown, L.L., Aron, A., Strong, G., & Mashek, D. (2010). Reward, addiction, and emotion regulation systems associated with rejection in love. *Journal of Neurophysiology, 104*(1), 51–60.

Fisher, H.E., Xu, X., Aron, A., & Brown, L.L. (2016). Intense, passionate, romantic love: A natural addiction? How the fields that investigate romance and substance abuse can inform each other. *Frontiers in Psychology, 7*, 687.

Fletcher, G.J., Simpson, J.A., Campbell, L., & Overall, N.C. (2015). Pair-bonding, romantic love, and evolution: The curious case of Homo sapiens. *Perspectives on Psychological Science, 10*(1), 20–36.

Jordan, H.W., & Howe, G. (1980). De Clerambault syndrome (erotomania): A review and case presentation. *Journal of the National Medical Association, 72*(10), 979–985.

Kawamichi, H., Sugawara, S.K., Hamano, Y.H., Makita, K., Matsunaga, M., Tanabe, H.C., ... Sadato, N. (2016). Being in a romantic relationship is associated with reduced gray matter density in striatum and increased subjective happiness. *Frontiers in Psychology, 7*, 1763.

Kross, E., Berman, M.G., Mischel, W., Smith, E.E., & Wager, T.D. (2011). Social rejection shares somatosensory representations with physical pain. *Proceedings of the National Academy of Sciences, 108*(15), 6270–6275.

Olojugba, C., de Silva, R., Kartsounis, L.D., Royan, L., & Carter, J. (2007). De Clerambault's syndrome (erotomania) as a presenting feature of fronto-temporal dementia and motor neurone disease (FTD-MND). *Behavioural Neurology, 18*(3), 193–195.

Song, H., Zou, Z., Kou, J., Liu, Y., Yang, L., Zilverstand, A., ... Zhang, X. (2015). Love-related changes in the brain: A resting-state functional magnetic resonance imaging study. *Frontiers in Human Neuroscience, 9*, 71.

Sturm, V.E., Yokoyama, J.S., Eckart, J.A., Zakrzewski, J., Rosen, H.J., Miller, B.L., ... Levenson, R.W. (2015). Damage to left frontal regulatory circuits produces greater emotional reactivity in frontotemporal dementia. *Cortex*, *64*, 55–67.

Xu, X., Brown, L., Aron, A., Cao, G., Feng, T., Acevedo, B., & Weng, X. (2012). Regional brain activity during early-stage intense romantic love predicted relationship outcomes after 40 months: An fMRI assessment. *Neuroscience Letters*, *526*(1), 33–38.

7 FROM SAFETY PINS TO SLEEPING BEAUTIES

American Psychiatric Association. (2013). *Diagnostic and Statistical Manual of Mental Disorders: DSM-5* (5th ed.). Washington, DC: American Psychiatric Association.

Ballard, J.G. (1973). *Crash*. London: Jonathan Cape.

Bianchi-Demicheli, F., Rollini, C., Lovblad, K., & Ortigue, S. (2010). 'Sleeping Beauty paraphilia': Deviant desire in the context of bodily self-image disturbance in a patient with a fronto-parietal traumatic brain injury. *Medical Science Monitor*, *16*(2), C15–C17.

Burns, J.M., & Swerdlow, R.H. (2003). Right orbitofrontal tumor with pedophilia symptom and constructional apraxia sign. *Archives of Neurology*, *60*(3), 437–440.

Devinsky, J., Sacks, O., & Devinsky, O. (2010). Klüver-Bucy syndrome, hypersexuality, and the law. *Neurocase*, *16*(2), 140–145.

Fumagalli, M., Pravettoni, G., & Priori, A. (2015). Pedophilia 30 years after a traumatic brain injury. *Neurological Sciences*, *36*(3), 481–482.

Mendez, M., & Shapira, J.S. (2011). Pedophilic behavior from brain disease. *Journal of Sexual Medicine*, *8*(4), 1092–1100.

Mitchell, W., Falconer, M., & Hill, D. (1954). Epilepsy with fetishism relieved by temporal lobectomy. *Lancet*, *264*(6839), 626–630.

Mohnke, S., Mueller, S., Amelung, T., Krüger, T.H., Ponseti, J., Schiffer, B., ... Walter, H. (2014). Brain alterations in paedophilia: A critical review. *Progress in Neurobiology*, *122*, 1–23.

Ortego, N., Miller, B.M., Itabashi, H., & Cummings, J.L. (1993). Altered sexual behavior with multiple sclerosis: A case report. *Neuropsychiatry, Neuropsychology, and Behavioral Neurology*, *6*(4), 260–264.

Sacks, O. (2019). Urge. In O. Sacks, *Everything In Its Place: First loves and last tales* (pp. 108–113). New York: Knopf.

Tenbergen, G., Wittfoth, M., Frieling, H., Ponseti, J., Walter, M., Walter, H., ... Kruger, T.H. (2015). The neurobiology and psychology of pedophilia: Recent advances and challenges. *Frontiers in Human Neuroscience*, *9*, 344.

8   SEX ON THE SPECTRUM

American Psychiatric Association. (2013). *Diagnostic and Statistical Manual of Mental Disorders: DSM-5* (5th ed.). Washington, DC: American Psychiatric Association.

American Psychological Association. (2018). Orgasmic reconditioning. *APA Dictionary of Psychology*, <https://dictionary.apa.org/orgasmic-reconditioning>.

Baird, A. (2009). *University challenge: A student with Asperger Syndrome shares his rage.* Paper presented at the Asia Pacific Autism Conference, 20–22 August, Sydney.

Barahona-Corrêa, J.B., & Filipe, C.N. (2016). A concise history of Asperger Syndrome: The short reign of a troublesome diagnosis. *Frontiers in Psychology*, 6, 2024

Chandradasa, M., & Champika, L. (2017). Zoophilia in an adolescent with high-functioning autism from Sri Lanka. *Australasian Psychiatry*, 25(5), 486–488.

Ecker, C. (2017). The neuroanatomy of autism spectrum disorder: An overview of structural neuroimaging findings and their translatability to the clinical setting. *Autism, 21*(1), 18–28.

Fernandes, L.C., Gillberg, C.I., Cederlund, M., Hagberg, B., Gillberg, C., & Billstedt, E. (2016). Aspects of sexuality in adolescents and adults diagnosed with autism spectrum disorders in childhood. *Journal of Autism and Developmental Disorders*, 46(9), 3155–3165.

Schöttle, D., Briken, P., Tüscher, O., & Turner, D. (2017). Sexuality in autism: Hypersexual and paraphilic behavior in women and men with high-functioning autism spectrum disorder. *Dialogues in Clinical Neuroscience*, 19(4), 381–393.

Silberman, S. (2015). *NeuroTribes: The legacy of autism and the future of neurodiversity.* London: Penguin.

van Rooij, D., Anagnostou, E., Arango, C., Auzias, G., Behrmann, M., Busatto, G.F., … Buitelaar, J.K. (2017). Cortical and subcortical brain morphometry differences between patients with autism spectrum disorder and healthy individuals across the lifespan: Results from the ENIGMA ASD working group. *American Journal of Psychiatry*, 175(4), 359–369.

9   PORN ON THE TRAIN (AND ON THE BRAIN)

American Psychiatric Association. (1987). *Diagnostic and Statistical Manual of Mental Disorders: DSM-III-R* (3rd rev. ed.). Washington, DC: American Psychiatric Association.

American Association of Sexuality Educators, Counselors and Therapists

[AASECT]. (2016). AASECT position on sex addiction. AASECT, <www.aasect.org/position-sex-addiction>.

Kraus, S.W., Krueger, R.B., Briken, P., First, M.B., Stein, D.J., Kaplan, M.S., ... Reed, G.M. (2018). Compulsive sexual behaviour disorder in the ICD-11. *World Psychiatry*, *17*(1), 109–110.

Kühn, S., & Gallinat, J. (2014). Brain structure and functional connectivity associated with pornography consumption: The brain on porn. *JAMA Psychiatry*, *71*(7), 827–834.

—— (2016). Neurobiological basis of hypersexuality. *International Review of Neurobiology*, *129*, 67–83.

Ley, D.J. (2012). *The Myth of Sex Addiction*. Lanham, MD: Rowman & Littlefield.

—— (2018). Most behavioral addictions get better on their own. *Psychology Today*, 23 April, <www.psychologytoday.com/us/blog/women-who-stray/201804/most-behavioral-addictions-get-better-their-own>.

Meurk, C., Carter, A., Partridge, B., Lucke, J., & Hall, W. (2014). How is acceptance of the brain disease model of addiction related to Australians' attitudes towards addicted individuals and treatments for addiction? *BMC Psychiatry*, *14*, 373.

Prause, N. (2017). Evaluate models of high-frequency sexual behaviors already. *Archives of Sexual Behavior*, *46*(8), 2269–2274.

Rush, B. (1812). *Medical inquiries and observations upon the diseases of the mind*. Philadelphia: Kimber & Richardson.

Satel, S., & Lilienfeld, S.O. (2013). Addiction and the brain-disease fallacy. In S. Satel & S.O. Lilienfeld (eds), *Brainwashed: The seductive appeal of mindless neuroscience* (pp. 49–72). New York: Basic Books.

Seok, J.-W., & Sohn, J.-H. (2015). Neural substrates of sexual desire in individuals with problematic hypersexual behavior. *Frontiers in Behavioral Neuroscience*, *9*, 321.

—— (2018). Gray matter deficits and altered resting-state connectivity in the superior temporal gyrus among individuals with problematic hypersexual behavior. *Brain Research*, *1684*, 30–39.

Sescousse, G., Caldú, X., Segura, B., & Dreher, J.C. (2013). Processing of primary and secondary rewards: A quantitative meta-analysis and review of human functional neuroimaging studies. *Neuroscience & Biobehavioral Reviews*, *37*(4), 681–696.

10 BLAME IT ON MY BRAIN

Bartlett, F., Hall, W., & Carter, A. (2013). Case and comment. Tasmania v Martin (No 2): Voluntariness and causation for criminal offending associated with treatment of Parkinson's disease. *Criminal Law Journal*, *37*(5), 330–341.

Blair, R.J., & Cipolotti, L. (2000). Impaired social response reversal: A case of 'acquired sociopathy'. *Brain*, *123*(6), 122–141.

Burns, J.M., & Swerdlow, R.H. (2003). Right orbitofrontal tumor with pedophilia symptom and constructional apraxia sign. *Archives of Neurology*, *60*(3), 437–440.

Carter, A., Ambermoon, P., & Hall, W.D. (2009). Drug-induced impulse control disorders: A prospectus for neuroethical analysis. *Neuroethics*, *4*(2), 91–102.

Damasio H, Grabowski T, Frank R, Galaburda AM, Damasio AR. (1994). The Return of Phineas Gage: Clues about the brain from the skull of a famous patient. *Science*, *264*(5162):1102–5.

Devinsky, J., Sacks, O., & Devinsky, O. (2010). Klüver-Bucy syndrome, hypersexuality, and the law. *Neurocase*, *16*(2), 140–145.

Eslinger, P.J., & Damasio, A.R. (1985). Severe disturbance of higher cognition after bilateral frontal ablation: Patient EVR. *Neurology*, *35*(12), 1731–1741.

Green, C. (2008). Paedophilia blamed on teacher's medication. *Independent*, 12 September, <www.independent.co.uk/news/uk/home-news/paedophilia-blamed-on-teachers-medication-927095.html>.

Hagan, K. (2014). Parkinson's disease sufferers win payout from Pfizer for drug linked to gambling, sex addiction. *Sydney Morning Herald*, 8 December, <www.smh.com.au/healthcare/parkinsons-disease-sufferers-win-payout-from-pfizer-for-drug-linked-to-gambling-sex-addiction-20141208-122jwk.html>.

Irvine, C. (2008). Parkinson's caused teacher's child porn habit, judge rules. *Telegraph*, 12 September, <www.telegraph.co.uk/news/uknews/2801663/Child-porn-habit-caused-by-Parkinsons.html>.

Judicial College of Victoria. (2018). *Victorian Sentencing Manual* (s. 10.9 – Mental impairment). Judicial College of Victoria, <www.judicialcollege.vic.edu.au/eManuals/VSM/toc.htm>.

Mobbs, D., Lau, H.C., Jones, O.D., & Frith, C.D. (2007). Law, responsibility, and the brain. *PLoS Biology*, *5*(4), e103.

Munhoz, R.P., Fabiani, G., Becker, N., & Teive, H.A. (2009). Increased frequency and range of sexual behavior in a patient with Parkinson's disease after use of pramipexole: A case report. *Journal of Sexual Medicine*, *6*(4), 1177–1180.

Neuroskeptic. (2009). Brain damage, pedophilia, and the law [blog post]. Neuroskeptic, 23 November, <http://blogs.discovermagazine.com/neuroskeptic/2009/11/23/brain-damage-pedophilia-and-the-law>.

New South Wales Department of Justice. (2019). Statewide justice advocacy for people with cognitive impairment [media release]. Department of

Justice, 5 April, <www.justice.nsw.gov.au/Pages/media-news/media-releases/2019/statewide-justice-advocacy-for-people-with-cognitive-development.aspx>.

Ortego, N., Miller, B.M., Itabashi, H., & Cummings, J.L. (1993). Altered sexual behavior with multiple sclerosis: A case report. *Neuropsychiatry, Neuropsychology, and Behavioral Neurology*, 6(4), 260–264.

Sacks, O. (2019). Urge. In O. Sacks, *Everything In Its Place: First loves and last tales* (pp. 108–113). New York: Knopf.

Sage, A. (2012). GSK ordered to pay Parkinson's sufferer for Requip side effects. *The Australian*, 29 November, <www.theaustralian.com.au/news/world/gsk-ordered-to-pay-parkinsons-sufferer-for-requip-side-effects/news-story/99c7685e012bd8758f4e25e913edc620>.

Sentencing Advisory Council (Vic.). (2017). Mental impairment and sentencing. Sentencing Advisory Council, <www.sentencingcouncil.vic.gov.au/about-sentencing/sentencing-process/mental-impairment>.

Student-turned-dominatrix awarded $1.5M after car accident left her with brain injury and a new personality. (2012). *National Post*, 29 January, <https://nationalpost.com/news/canada/student-turned-dominatrix-awarded-1-5m-after-car-accident-left-her-with-brain-injury-and-a-new-personality>.

Williams, R. (2012). Parkinson's sufferer wins six figure payout from GlaxoSmithKline over drug that turned him into a 'gay sex and gambling addict'. *Independent*, 29 November, <www.independent.co.uk/news/world/europe/parkinsons-sufferer-wins-six-figure-payout-from-glaxosmithkline-over-drug-that-turned-him-into-a-gay-8368600.html>.

ACKNOWLEDGMENTS

To my publisher Phillipa McGuinness, thank you for the opportunity to make this book a reality. I always left our lunch meetings confident and inspired to continue, and I have greatly valued your warm and inspiring mentorship throughout this journey. To Emma Driver, copyeditor extraordinaire, your thought-provoking queries and efficiency made the process fun. To the late Oliver Sacks, whose books inspired me to become a clinical neuropsychologist and to write a book myself. To my PhD supervisors Sarah Wilson, David Reutens and Peter Bladin, thank you for introducing me to this fascinating topic and for facilitating my PhD, which was the seed of this book. To my interviewees Till Amelung, John Christie, David Ley, Nicole Prause, Brian Russman and Vanessa Thompson, thank you for your time and insights, which made an important contribution to this book, and thanks to Justin Lehmiller and Adrian Carter for pointing me in the direction of relevant literature for Chapters 9 and 10 respectively. To George, thank you for your friendship and for allowing me to discuss your life. To my late father, Rob, who no doubt had an influence on my drive to understand the brain. To my parents Gabriele and Drew, your love and endless encouragement to pursue my dreams have enabled me to do just that. Thank you to my husband, Tim, for your calm patience in listening to me talk about this book and my career fantasies over the years, and for letting me disappear for hours on weekends to write. To my children Louis, Clara and Theo, I hope that once you are old enough to read this book you will understand why I was so excited to write it. Finally, to my patients, thank you for your candid discussions about the intimate details of your life. It is a privilege to have met you.

INDEX